COMPARATIVE PHYSIOLOGY

FUNCTIONAL ASPECTS OF STRUCTURAL MATERIALS

COMPARATIVE PHYSIOLOGY
FUNCTIONAL ASPECTS
OF STRUCTURAL MATERIALS

Proceedings of the International Conference on
Comparative Physiology, Ascona, 1974

Editors:

L. Bolis
University of Rome, Rome, Italy

S.H.P. Maddrell
*Agricultural Research Council,
Cambridge, Great Britain*

K. Schmidt-Nielsen
Duke University, Durham, U.S.A.

1975

NORTH-HOLLAND PUBLISHING COMPANY — AMSTERDAM · OXFORD
AMERICAN ELSEVIER PUBLISHING COMPANY, INC. — NEW YORK

Library of Congress Catalog Card Number 75-1765

ISBN North-Holland 0-7204-4532-9
ISBN American Elsevier 0-444-10944-7

9-13-77

NOV 23 1977

Publishers:

NORTH-HOLLAND PUBLISHING COMPANY — AMSTERDAM
NORTH-HOLLAND PUBLISHING COMPANY, LTD. — OXFORD

Sole distributors for the U.S.A. and Canada:
AMERICAN ELSEVIER PUBLISHING COMPANY, INC.
52 VANDERBILT AVENUE
NEW YORK, N.Y. 10017

PRINTED IN THE NETHERLANDS

Preface

This book, like its predecessor, shows what can be gained from studying a variety of organisms. The reader will find here treatments of such apparently diverse topics as the mechanisms whereby bacteria beat their flagella, locusts tan their cuticle, and the energy cost of contraction of chicken muscle. The point of including such an array is not merely to illustrate differences, but also, of course, to uncover those physiological mechanisms which are fundamental to animals — in short, to find out how animals work. Just how far progress has gone along these lines in the fields dealt with in this volume, the reader must judge for himself.

One of our main objects in holding Congresses on Comparative Physiology is to bring together contributions on the same topics from workers in different disciplines. Part I of this volume is an excellent example; it deals with the role of materials in organisms, and contains papers from biophysicists and engineers as well as from biologists. Many of us present at the Congress found this to be a most fertile mixture and came away having learnt a great deal. We hope that publishing these papers together here will have the same happy result.

Simon Maddrell

List of contributors

Amos, W.B.
Andersen, S.O.
Anner, B.
Beamish, F.W.J.
Biggs, W.D.
Bolis, L.
Caldwell, P.C.
Chowrashi, P.K.
De Peyer, J.E.
Evans, E.A.
Ferrero, J.
Fletcher, C.R.
Goldspink, G.
Gordon, J.E.
Gordon, M.S.
Haselgrove, J.C.
Huggel, H.
Jirounek, P.

Keynes, R.D.
Kilburn, D.G.
Lett, P.F.K.P.
Molinaro, M.
Morley, M.P.
Niimi, A.J.
Pepe, F.A.
Pringle, J.W.S.
Rankin, J.C.
Routledge, L.M.
Straub, R.W.
Wachsberger, P.R.
Wainwright, S.A.
Weis-Fogh, T.
Wilbrandt, W.
Yensen, J.
Zani, B.

Contents

Preface V
List of contributors VII

Part 1 *Mechanical function of structural materials*

The role of materials in organisms
    S.A. Wainwright . . . . 3—8
Two-dimensional, hyperelastic materials
    E.A. Evans . . . . 9—24
Time dependent properties — their measurement and meaning
    W.D. Biggs . . . . .25—41
Structural proteins in relation to mechanical function: Cross-links in
    insect cuticle
    S.O. Andersen . . . . 43—48
Mechanical instabilities in biological membranes
    J.E. Gordon . . . . 49—57

Part 2 *Contractile mechanisms*

Bacterial flagella: structure and function
    L.M. Routledge . . . . 61—73
Bioelectric control of ciliary activity
    J.E. De Peyer and H. Huggel . . . . 75—82
Principles of contraction in the spasmoneme of vorticellids. A new
contractile system
    T. Weis-Fogh . . . . 83—98
Structure and protein composition of the spasmoneme
    W.B. Amos . . . . 99—104
Myosin filaments of skeletal and uterine muscle
    F.A. Pepe, P.K. Chowrashi and P.R. Wachsberger . . . . . . .105—119
Molecular differentiation and myoblast fusion during myogenesis in
culture
    B. Zani and M. Molinaro . . . . .121—126
Structural changes in smooth and striated muscle during contraction
    J.C. Haselgrove .127—138
Insect fibrillar muscle and the problem of contractility
    J.W.S. Pringle .139—152

X

Part 3 *Energy demands on animals*

The energy cost of active transport
R.D. Keynes . . . . .155—159
The energetics of ionic regulation
C.R. Fletcher .161—172
Biochemical energetics of fast and slow muscles
G. Goldspink .173—185
Bioenergetics of teleost fishes: Environmental influences
F.W.H. Beamish, A.J. Niimi and P.F.K.P. Lett. . . . . . . .187—209
Effects of temperature and pressure on the oxidative metabolism of
fish muscle
M.S. Gordon .211—222
Adrenergic control of blood flow through fish gills: Environmental
implications
L. Bolis and J.C. Rankin . . . . .223—233
Energy costs of ion pumping and the drain on the cell's metabolism
D.G. Kilburn, M.P. Morley and J. Yensen . . . . .235—241
Energy utilization for active transport by the squid giant axon
P.C; Caldwell 243- 248
Transport of inorganic phosphates across nerve membranes
R.W. Straub, B. Anner, J. Ferrero and P. Jirounek . . . . . .249—257
Energetics of sugar transport
W. Wilbrandt .259—265

Subject index .267

PART 1

MECHANICAL FUNCTION OF STRUCTURAL MATERIALS

Comparative Physiology — Functional Aspects of Structural Materials
Eds L. Bolis, H.P. Maddrell and K. Schmidt-Nielsen
© North-Holland Publishing Company — 1975 — Amsterdam

The role of materials in organisms

S.A. WAINWRIGHT

Zoology Department Duke University, Durham, N.C. 27706 (U.S.A.)

INTRODUCTION

Until quite recently our knowledge of mechanical properties of biological structural materials has been characterised by a small number of studies by a small number of workers each of whom is a biologist trying to learn enough about polymer physics or the strength of engineering materials to make an intelligent investigation of a biomaterial. The result is a number of rather isolated studies of materials seen only in the wider context of man-made materials. It is delightfully clear from this little symposium that some general statements about biological materials and their role in organisms can now be made and that directions of future research are indicated. I would like to point out some generalizations and directions in three areas of study: (1) biomaterials as materials, (2) the formation of biomaterials and (3) biomaterials as constituents of structural systems.

BIOMATERIALS AS MATERIALS

The alloys and reinforced concrete of engineers are Hookean solids; among well-known man-made polymers we know about crystalline and strain-crystallizing polymers, viscous polymers and rubberlike polymers. Biomaterials have had to be described in comparison to these because no other theoretical body of knowledge was available. Biologists were learning from physical scientists, but it was never supposed that biologists might have something to offer in return.

Using the approach of continuum mechanics, E.A. Evans has shown that the hyperelastic membranes of red blood cell membranes are not typical elastic solids but rather they behave like two-dimensional materials. This is a new concept in materials: we know about two-dimensional liquid films, but membrane materials are demonstrably solids. Dr. Evans continues to describe their behavior and he is also looking into the membranes of other cell types to determine the extent to which the concept is useful there as well. Of particular interest is the molecular mechanism by which the membrane material undergoes the large, elastically reversible strains. This subject will also yield new concepts for materials scientists.

W.D. Biggs has set out the context of time-dependent properties wherein biologists may work in order to reveal the general features of biomaterials as well as the bases of their incredible diversity. He suggests that by knowing the stress, the strain and the strain rate of a sample, a mechanical equation of state can be created that will allow the comparison of diverse materials without regard to their recent mechanical history. This latter consideration has been a major stumbling block in several biomechanics laboratories (e.g. Fung [1]). Happily, Biggs has shown us which tests to perform and he has suggested a basic set of terms that, if followed by all, will greatly facilitate communication among biologists and materials scientists. It is abundantly clear at this Conference that the communication of simple mechanical concepts between scientists who have been raised in different specialties is indeed a problem. Unlike many social problems, this one is easily solved.

In the work on elastin with J.M. Gosline, T. Weis-Fogh shows the world a previously undescribed mechanism for the elastic return from large deformation: that of using surface energy at the interface between aqueous and non-aqueous phases in combination with the entropic energy of conformation change in randomly coiled molecules to effect elastic return. It is significant that this biological mechanism is more complex than the previously known mechanism for long range elasticity: for the most part, biological materials are much more complex than those studied by polymer scientists. Similarly, in the Symposium on Contractile Materials in this volume, Weis-Fogh and Amos introduce an entirely new contractile material with a new mechanism for contraction.

The study of biomaterials has come of age: materials scientists can now learn from biologists.

SYNTHESIS OF BIOMATERIALS

Biologists are painfully aware that the articulated skeleton of the adult arthropod or vertebrate and the stem of an erect angiosperm arose each in a different manner as an integrated set of products made by cells. A tree grows by adding material to the outside of what went before. Bones grow by continual remodelling, replacement and change in composition, properties and shape. Insect cuticle is formed, noncrosslinked and folded before the molt. After the molt it is hardened to a shape whose surface area does not grow again until the next molt but which continues to grow in thickness by adding layers daily on the inside.

Polymer scientists I have known quail at the complexity of biological materials. To ask such folk to create a complex system of complex materials that must be able to grow in one of these modes continuously and without interruption of mechanical integrity or function is to invite his total relapse. To ask a biologist how any living system accomplishes this invites a mystified shrug. S.O. Andersen's work is the first to actually describe the qualitative differences in the hardening (crosslinking) components in cuticle from various parts of the insect body and their quantitative changes during development. It is comforting that he concludes that control of crosslinking resides in the epidermal cells immediately underlying and secreting the cuticle. However, development involves the entire organism and while the properties of cuticles are controlled by the secretory activities of adjacent cells, Maddrell [2], for example, has shown that the control of these epidermal cells may reside in a neurosecretory pathway that originates far from the area of change in cuticular properties. The development of structural systems and their constituent materials remains an area where each morsel of research is lonesome, but work like Andersen's shows us how understanding of the integrated system can be achieved.

MATERIALS AS CONSTITUENTS OF SYSTEMS

Just as the materials made by man must function as constituents of automobiles, cathedrals, windmills and soccer boots, the overwhelmingly important thing about biomaterials is that they are functional parts of whole organisms. Looked at as solutions to design problems in comparison with man-made systems, biological systems are, on the whole, less rigid, less strong and are capable of less rapid and less powerful movements than the best of man's products. As J.E. Gordon says, organisms go in for low structure loading coefficients. However, because of the peculiarities of their modes of synthesis, organisms are continually replacing and repairing their materials, and they appear functionally to be immune to fatigue. Also, biomaterials are formed at "room temperature" and pressure.

If absolute strength and stiffness are down, how do organisms survive storms and waves and how do they swim, run and fly? Most forces created by organisms or met with by them in their habitat are not impact forces. Therefore, if the forces themselves build up over finite periods of time and strain rates are low, it is possible to have materials and systems that absorb high strain energy. One of the mechanisms that ensure low strain rates and allows absorption of high strain energy is that organisms and their materials accept large strains. This is the point of Gordon's paper where he points out that strains in soft tissues are independent of stress up to 60% strain or more. This is an immensely important generality: it describes the predominant mechanical feature of all soft tissues in all organisms. Since all organisms are not identical, one can conclude that a very big piece of the business of evolution has been the designing of different materials to absorb greater strain energy.

In fact, organisms bend [3]. Rye plants and coconut palms bend with the wind. Feathers and wings and fins bend with fluid dynamic force. Suspension-feeding sea fans, crinoids and hydroids that are attached to the sea floor bend with the currents. Because bending is a complex type of strain that puts part of the system in tension and part in compression, one might expect the number of possible combinations of materials and geometries to be large. In fact, it appears that the gross shapes of animals that allow us to divide them into more than twenty phyla are actually the expressions of the solutions to this basic mechanical problem. Each phylum is characterized by a particular array of compression-resisting materials and bend-resisting systems that is unique. Since representatives of all these groups are extant, we must assume that they are all successful and that none is perfect or even substantially more effective than any other. A very modest selective advantage gives a species dominance in very few generations: it has taken millenia to create these diverse groups by selection.

Gordon modestly treats soft tissues as black boxes and does not mention how these tissues accomodate to large deformations. Aside from the conformational changes and entanglements of very long, randomly-coiled molecules that are known to be the basis of much hyperelasticity in soft tissues (e.g. [4,5]), there is a wide variety of cylindrical animals and plants and their parts that have evolved a variation of the plaited finger trap. This gadget is a tube whose inner diameter is roughly finger-sized. It is made of right and left-hand helically wound fibers. If you put a finger into each end and then try to pull them out, friction makes the tube stick to the fingers and the tube is stretched. For every unit of stretch in the tube, the diameter shrinks accordingly and the fingers are trapped.

Body walls of many wormlike animals [6] and the cell walls of most higher plants [7] have such a crossed-helical structure of high modulus fibrous material in a visco-elastic matrix and they behave as expected. This arrangement of fibers in a composite material allows the organism or part thereof to undergo large deformations. This feature may be a reason that these fibrous systems are helically wound rather than rectilinearly wound throughout the plant and animal kingdoms.

TIME-DEPENDENCE

The other aspect of biomaterials that makes them good energy absorbing materials is their visco-elasticity, that is their ability to distribute stresses according to the rate at which they are strained. Biggs tells us why the relaxation and retardation times and the creep and creep recovery are important in predicting the molecular structure of polymeric materials. A material may exist in two organisms that differ in their rates of movement: this difference may rely as heavily on the relaxation time of the skeletal materials involved as it does on the physiology of neuromuscular response. Alexander [8,9] thus showed that while the sea anemone *Metridium* and the medusa *Aequorea* both have a collagenous connective tissue called mesoglea that controls their body shapes and permits the slow peristaltic swaying of the anemone and the one-per-second swimming beats of the medusa, the behaviors of the two animals are predictable from a knowledge of the distribution of relaxation times of the two mesogleas.

Currently M.K. Glasser is comparing the time-dependent properties and behavior of *Metridium* that lives in calm water and is capable of large changes in body shape and size with those of *Anthopleura xanthogrammica*, an anemone that lives in wave-beaten surge channels. The volume fraction of mesogleal collagen is higher in *Anthopleura* than in *Metridium*: this contributes to the much greater tensile modulus of the former. The much longer mesogleal retardation time of *Anthopleura* is probably due to greater length between cross-links of random-coiled molecules in the intercollagenous matrix and the concentration of these molecules. The important thing is that *Anthopleura* mesoglea is able to accommodate to the near-impact of a large wave (duration 8 s) and can elastically recover its deformation before the next wave hits (8 s). In a similar situation, *Metridium* required on the order of 20 s to recover.

In summary of this point, we find that via the time-dependent properties of their constituent materials, organisms are tuned to the forces and the range of strain rates the organisms meet in their habitats. More simply, organisms are as finely adapted to mechanical features of their environment as they are to light intensity, O_2 tension, temperature, etc.

Finally, there is a new functional description of the supportive system of a seastar, *Asterias forbesi*, based on observation of morphology and the mechanical properties that shows all of these points in one example. The supportive system of the seastar arm contains a set of interacting calcareous ossicles that are about 1.0 mm in their largest dimension. J.P. Eylers has shown that wherever two ossicles of the aboral body wall meet, they meet at a flat bearing surface and that they are connected together by a muscle running parallel with a bundle of collagen fibres. The ossicles form a polygonal array over the entire upper surface of the arm. The arm can be bent and twisted by the seastar into any sort of coiled shape. As the arm is bent, ossicles remain in contact and slide on their bearing surfaces. Characteristic of seastars and most other Echinodermata (the phylum of seastars, sea cucumbers, crinoids, brittle stars and sea urchins), the movements are very slow by our standards. What is interesting is that the seastar can, by simply contracting the muscles between ossicles, stop the movement and freeze the arm into incredible rigidity no matter what shape it has. Thus, by combining multiple tiny compression elements (ossicles) in a tension web of muscle and collagen and by acting slowly but with great force, the seastar has built a beam with short struts whose shape is continually variable and which can be "hardened" at will in any shape. This allows the seastar to open oysters by force in awkward places and, once the seastar is locked into a pocket in a rock, it resists being collected by predator or biologist without either it or the rock being destroyed. As a structural system compared to the trees, vertebrates and arthopods we know, it is a drastic and exciting innovation in skeletal design.

Summarizing again, the understanding of structure and mechanical function of biomaterials provides insight that greatly facilitates our understanding of whole organisms, their lifestyles and their evolution.

REFERENCES

1. Fung, Y.C.B. (1967) Elasticity of soft tissues in simple elongation. Am. J. Physiol. 213, 1532—1544.
2. Maddrell, S.H.P. (1966) Nervous control of the mechanical properties of the abdominal wall at feeding in *Rhodnius*. J. Exp. Biol. 44, 59—68.
3. Wainwright, S.A., Biggs, W.D., Currey, J.D. and Gosline, J.M. (1975) Mechanical Design in Organisms. Edward Arnold, London.
4. Gosline, J.M. (1971) Connective tissue mechanics of *Metridium senile*. I. Structural and compositional aspects. J. Exp. Biol. 55, 763—774.
5. Gosline, J.M. (1971) Connective tissue mechanics of *Metridium senile*. II. Visco-elastic properties and macromolecular model. J. Exp. Biol. 55, 775—795.
6. Clark, R.B. (1964) Dynamics in Metazoan Evolution. Oxford University Press, London.
7. Roelofsen, P.A. (1959). The Plant Cell Wall. Gebr. Borntraeger, Berlin.
8. Alexander, R.M. (1962) Viscoelastic properties of the body wall of anemones. J. Exp. Biol. 39, 373—386.
9. Alexander, R.M. (1964) Viscoelastic properties of the mesogloea of jellyfish. J. Exp. Biol. 41, 363—369.

Comparative Physiology — Functional Aspects of Structural Materials
Eds L. Bolis, H.P. Maddrell and K. Schmidt-Nielsen
© North-Holland Publishing Company — 1975 — Amsterdam

Two-dimensional, hyperelastic materials

E.A. EVANS

Departments of Biomedical Engineering and Surgery, Duke University, Durham, N.C. (U.S.A.)

INTRODUCTION

The motivation for measuring properties of two-dimensional materials has come from the study of mechanical experiments on biological plasma membranes and lipid bilayer replicas [1→3]. From these experiments, it is apparent that the membrane material behaves anisotropically. In other words, stresses in the plane of the membrane are not coupled to the direction normal to the membrane, i.e. cannot change the thickness of the surface. From an ultrastructural viewpoint, this is not surprising because the material is a composite of molecular monolayers: a continuum in two dimensions with molecular character in the third. By contrast, a three-dimensional, isotropic material surface changes thickness in response to area change.

First impressions cause one to consider "soap bubble" mechanics. However, it must be recalled that "soap bubble" mechanics is based on the premise of a free, liquid interface: surface can be "created" and "destroyed" from liquid molecules. In addition, the law of Laplace (which characterizes "soap bubble" mechanics) presumes that the tensions (stress resultants) in the plane of the material are isotropic (2-D) which is only true in general for a liquid surface. On the contrary, the membrane materials observed (and undoubtedly there are other structures not recognized here where 2-D materials are present) exhibit "elastic", solid behavior and are not free surfaces; the total number of surface molecules is fixed.

Therefore, the requirement is to develop constituitive relations for a two-dimensional material surface and composite layers. In this paper, the elastic, solid behavior will be discussed. The results will be valid for nonlinear, nonhomogeneous surfaces and finite deformations. Because the surface elements have fixed thickness, perhaps only a single molecule, the forces may be considered as distributed per unit length on the sides of the surface element. The concept of force per unit area on the element sides (stresses) is not viable here; therefore, the label tension is chosen. Consequently, tension-strain laws will be developed as opposed to stress-strain laws; elastic and material moduli will have units of dynes/cm rather than dynes/cm^2. Pseudo-equivalent moduli can be obtained by dividing by the material thickness; however, this confuses the issue and is meaningless for single, molecular layers.

In the initial mathematical formulation of the next section, a convenient "cataloging" system will be used to keep track of the tension and strain components: the two-dimensional, cartesian tensor. Certain properties of these matrices, their invariants, will be used when the choice of local coordinates become unimportant, i.e. two-dimensional isotropy. It is suggested that the non-mathematical reader carefully investigate Eqns (1), (2), (6), (7), (8), and from (12) on, and not be distracted by the intermediate details.

The specific application will consider a highly extensible (hyperelastic) surface with small compressibility. Plasma membranes of biological cells and lipid bilayers constructed in vitro appear to belong to this class of materials. In addition, bending moments and resistance for two-dimensional material composites are developed and discussed for the first time.

TENSION-STRAIN LAW FOR FINITE DEFORMATION OF A TWO DIMENSIONAL MATERIAL

In this section, the tension-strain law for a two-dimensional material will be established similar to general three-dimensional developments (e.g. [4]). The tension-strain law is obtained from the partial derivatives of the elastic free energy density of the material, taken with respect to the material strains. The elastic free energy density is free energy per unit surface area (dynes $\cdot$ cm $\cdot$ cm^{-2}). The relationship will involve the use of extension ratios. For finite strains, it is necessary to make the distinction between Eulerian and Lagrangian strains. The former are based on the metric of the final state and the latter are relative to the metric of the initial state. Here, the Lagrangian strains will be used.

The principal extension ratios are defined by,

$$\lambda_1 = \frac{d\bar{x}_1}{dx_1}$$

$$\lambda_2 = \frac{d\bar{x}_2}{dx_2} \tag{1}$$

where the metric for the initial state is,

$$ds = \sqrt{dx_1^2 + dx_2^2}$$

and the final state,

$$d\bar{s} = \sqrt{dx_1^2 + dx_2^2}$$

dx_i, $d\bar{x}_i$ are the initial and deformed dimensions of a material element (i = 1, 2).

Using Eqn (1), the principal Lagrangian strains are given by,

$$\epsilon_1 = \tfrac{1}{2}(\lambda_1^2 - 1)$$

$$\epsilon_2 = \tfrac{1}{2}(\lambda_2^2 - 1) \tag{2}$$

Because a two-dimensional material is being considered, it is convenient to introduce a strain tensor that represents area change. This tensor is called Finger's strain tensor and is defined by,

$$\beta_{ij} = \delta_{ij} + 2\epsilon_{ij} \tag{3}$$

where δ_{ij} is the unit matrix and ϵ_{ij}, the Lagrangian strain tensor. The general Lagrangian strain tensor is,

$$\epsilon_{ij} = \frac{1}{2}\left(\frac{\partial \bar{x}_k}{\partial x_j}\frac{\partial \bar{x}_k}{\partial x_i}\right) - \delta_{ij} \tag{4}$$

(The summation convention

$$\frac{\partial \bar{x}_k}{\partial x_j}\frac{\partial \bar{x}_k}{\partial x_i} \equiv \sum_{k=1}^{2} \frac{\partial \bar{x}_k}{\partial x_j}\frac{\partial \bar{x}_k}{\partial x_i}$$

will be employed throughout the paper.)

The invariants of the two-dimensional tensor are,

$$B_1 = \lambda_1^2 + \lambda_2^2 = \beta_{ii}$$

$$B_2 = -\lambda_1^2\lambda_2^2 = \tfrac{1}{2}(\beta_{ij}\beta_{ji} - \beta_{ii}\beta_{jj}) \tag{5}$$

The second invariant is minus the area ratio squared.

For isothermal deformations, the change in free energy of the material is equal to the work of applied forces on the material.

$$(\delta E) = (\delta W) \tag{6}$$

In general, the free energy change of the material results from two contributions: internal energy change (molecular interaction) and the negative entropy change (configurational order) times temperature.

$$(\delta E) = (\delta U) - T(\delta S) \tag{7}$$

Relative to Lagrangian variables, the material tensions (in the undeformed coordinate system) are given by the partial derivatives of the elastic free energy density,

$$\overline{S}_{ij} = \frac{\partial E}{\partial \epsilon_{ij}} = 2 \frac{\partial E}{\partial \beta_{ij}} \tag{8}$$

$\overline{S}_{ij}$ is the Kirchoff tension tensor. However, while it is useful to use Lagrangian strains (because of their relation to the undeformed coordinate system), it is preferable to use Eulerian tensions because mechanical force equilibrium is determined in the deformed system. The Eulerian tension tensor is obtained from the Kirchoff tension tensor by the following transformation,

$$T_{ij} = \frac{\overline{\rho}}{\rho} \frac{\delta \overline{x}_i}{\delta x_p} \frac{\delta \overline{x}_j}{\delta x_q} S_{pq} \tag{9}$$

where ρ/ρ is the material density ratio of initial to final state densities (mass/area).

At this point, it is assumed that the material is isotropic in the plane of the surface, however, no assumption is made about homogeneity. With isotropy in two dimensions, the elastic free energy density E is only a function of the two invariants of the strain tensor because E is independent of coordinate rotation.

$$\frac{\partial E}{\partial \beta_{ij}} = \frac{\partial E}{\partial B_1} \frac{\partial B_1}{\partial \beta_{ij}} + \frac{\partial E}{\partial B_2} \frac{\partial B_2}{\partial \beta_{ij}} \tag{10}$$

Using eqns (5), (8), (9) and (10),

$$T_{ij} = 2 \frac{\overline{\rho}}{\rho} \left[B_2 \frac{\partial E}{\partial B_2} \delta_{ij} + \frac{\partial E}{\partial B_1} \beta_{ij} \right] \tag{11}$$

Eqn (11) is the tension-strain law for the two dimensional isotropic material in terms of Finger's strain tensor and the associated invariants. In order to convey physical insight into the material relation, the invariants B_2 and B_1 are interchanged with the fractional area change α and the quadratic variation of the extension ratios Δ.

$$\alpha \equiv \lambda_1 \lambda_2 - 1$$

$$\Delta \equiv \lambda_1^2 + \lambda_2^2 - 2 \tag{12}$$

Eqn (11) becomes,

$$T_{ij} = \left(\frac{\partial E}{\partial \alpha} + \frac{2}{\lambda_1 \lambda_2} \frac{\partial E}{\partial \Delta}\right) \delta_{ij} + \frac{4}{\lambda_1 \lambda_2} \frac{\partial E}{\partial \Delta} \epsilon_{ij} \qquad (13)$$

where the Lagrangian strain tensor has been employed. The first term on the right hand side of Eqn (12) represents the 2-D isotropic tension as illustrated in Fig. 1. The second term represents the anisotropic tension illustrated for uniaxial extension and simple shear in Fig. 1.

The coefficient of the first term essentially represents the resistance to area dilation and surface compressibility (2-D). The coefficient of the second

Two dimensional
Compressibility
(Area dilatation)

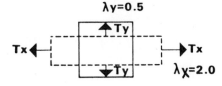

Simple extension
(Two dimensional
incompressibility)

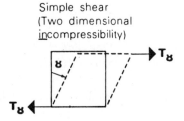

Simple shear
(Two dimensional
incompressibility)

Fig. 1. Illustration of two-dimensional material behavior.

term is the shear modulus or modulus of rigidity for large extensions at constant area. For a two-dimensional liquid surface, the partial derivative, $\partial E/\partial \Delta$, is zero. The surface tensions are isotropic,

$$T_{ij} = \frac{\partial E}{\partial \alpha} \delta_{ij} \tag{14}$$

However, additional terms representing the resistance to strain rate (viscous dissipation) would be required. For example,

$$T_{ij} = \frac{\partial E}{\partial \alpha} \delta_{ij} + \nu \dot{\epsilon}_{pq} \frac{\partial x_p}{\partial \bar{x}_i} \frac{\partial x_q}{\partial \bar{x}_j} \tag{15}$$

where $\dot{\epsilon}_{ij}$ is the strain rate tensor and ν is a coefficient of viscosity.

From Eqn 13, it is recognized that if the initial state is free of tension then,

$$\frac{\partial E}{\partial \alpha} + 2 \frac{\partial E}{\partial \Delta} = 0 \tag{16}$$

for $\alpha = \Delta = 0$.

FREE ENERGY COMPOSITION OF A HYPERELASTIC SURFACE WITH SMALL COMPRESSIBILITY

For a surface with small compressibility,

$$\lambda_1 \lambda_2 \simeq 1$$

$$\alpha \ll 1$$

$$\Delta \simeq \lambda_1^2 + \lambda_1^{-2} - 2 \tag{17}$$

A free energy model for such a material is given by,

$$E \simeq \gamma(1 + \alpha) + (\tfrac{1}{2} K_B \alpha^2 - \tfrac{1}{2} \mu_1 \alpha) + \sum_{n=1}^{\infty} \frac{\mu_n \Delta^n}{4} \tag{18}$$

The parameter γ is the surface chemical interaction free energy density (interfacial surface tension). K_B is the area compressibility modulus: "bulk" modulus (because the thickness is fixed). It is important to recognize that K_B may be different for area compression than for area dilation. The elements μ_n are coefficients in the non-linear shear modulus expansion. The term $\mu_1 \alpha/2$ is included to satisfy the Condition (16) in the relaxed state.

Three physico-chemical phenomena contribute to the free energy density Eqn (18):

(1) Surface constriction due to the chemical interaction with the adjacent materials (e.g. surface tension arising from immiscibility in surrounding fluids).

(2) Surface pressure resisting constriction or cohesive forces resisting expansion (two-dimensional analog of a liquid).

(3) Surface rigidity resisting extension and shear (exclusively property exhibited by solids). Using Eqns (13), (17) and (18), the tension-strain law is determined.

$$T_{ij} = \left(\gamma + K_B \alpha + \sum_{n=2}^{\infty} \frac{n\mu_n \Delta^{n-1}}{2} \right) \delta_{ij} + \left(\sum_{n=1}^{\infty} n\mu_n \Delta^{n-1} \right) \epsilon_{ij}$$

$$T_{ij} = (\gamma + K_B \alpha)\delta_{ij} + \mu\epsilon_{ij}$$

Only a single term for shear modulus has been retained in the latter expression.

In the equilibrium state, the material has internal tension determined by the balance of interfacial tension γ against compressibility (there is also a small contribution from $\mu\epsilon_{ij}$, but $\epsilon_{ij} \simeq 0$).

$$\gamma \simeq -K_B \alpha \tag{20}$$

This is valid for solid and liquid surfaces alike; in order to distinguish between the two, the shear modulus μ must be determined:

$\mu > 0$ for a solid

$\mu = 0$ for a liquid

Again, as discussed before, the viscoelastic behavior (for a linear viscoelastic solid) can be in coorporated into Eqn (19),

$$T_{ij} = (\gamma + K_B \alpha)\delta_{ij} + \mu\epsilon_{ij} + \nu\dot{\epsilon}_{pq} \frac{\partial x_p}{\partial \bar{x}_i} \frac{\partial x_q}{\partial \bar{x}_j} \tag{21}$$

Obviously no material is truely incompressible. However, from a practical point of view, if the bulk modulus of the material is much greater than the shear modulus, then the material behaves as if it is incompressible. The condition of incompressibility for a two-dimensional material is given by,

$$\lambda_1 \lambda_2 \equiv 1$$

For a two-dimensional, incompressible material,

$$K_B \gg \mu$$

and the isotropic term in Eqn (19) can be replaced by an isotropic tension, $-P_m$, a type of "hydrostatic" tension.

$$T_{ij} = -P_m \delta_{ij} + \mu \epsilon_{ij}$$

P_m is the locally isotropic tension that represents energy storage when resisting uniform dilation or compression of the surface area.

The nearly fixed surface area as opposed to a "free" surface yields entirely different mechanical behavior. Note that the interfacial tension, γ, is balanced by internal tensions in the material and the surface does not change shape to minimize the free energy (except the small value α). On the contrary, the "free" surface under the action of interfacial surface tension can exist only in the spherical state at equilibrium and must be balanced by external, normal stress.

MEMBRANE COMPOSITE STRUCTURES

Biological membranes have been characterized as possessing two, apparently opposite properties: fluidity and elastic rigidity. Strong evidence exists supporting both [3]. However, a composite membrane system made up of a structural element plus a superficial "liquid" sealer satisfies both sets of experiments and extensive ultrastructural observations. Fig. 2 illustrates a possible arrangement with an outer liquid, bimolecular layer and an inner "network" of structural material. Provided the surfaces maintain their two-dimensional character, the free energy densities are additive as shown in Fig. 2. Therefore, the tension-strain laws are the total of all constituent layers. This is true even if the layers rigidly connect. However, the multiplicity of layers introduces a new property in general to the composite: bending resistance.

The bending resistance for a monomolecular layer is essentially zero for radii of curvature much greater than molecular dimensions which is almost always the case. In other words, no resisting moments can be generated at the molecular bonds until the radii of curvature are very small. Therefore, if the layers are not connected, then no bending resistance exists. However, if the layers are rigidly connected, introducing curvature will produce tension in the "upper" surface or compression in the "lower" surface resulting in a bending moment.

MEMBRANE MATERIAL BEHAVIOR

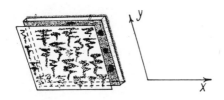

FREE ENERGY COMPOSITION FOR EACH LAYER

	Surface Free Energy Density (Interfacial Surface Tension)	Surface Compressibility (Eq'n. of State)	Surface Rigidity (Shear Resistance)
Layer 1	$\gamma_1\ A_1$	$K_1(A_1-A_1^\circ)^2/A_1^\circ$	$\frac{\mu}{4}(\lambda_x^2+\lambda_y^2-2)A_1$
Layer 2	$\gamma_2\ A_2$	$K_2(A_2-A_2^\circ)^2/A_2^\circ$	0
Layer 3	$\gamma_3\ A_3$	$K_3(A_3-A_3^\circ)^2/A_3^\circ$	0

For Thin Membrane (also requires $A_1^\circ = A_2^\circ = A_3^\circ$)

$$F_{TOTAL} = \gamma_T A + K_T\frac{(A-A^\circ)^2}{A^\circ}+\frac{\mu}{4}(\lambda_x^2 + \lambda_y^2 -2)A$$

Fig. 2. Free energy partition of a membrane composite (schematically illustrated as a lipid bilayer with underlying "network").

BENDING MOMENTS AND RIGIDITY CONSTANTS

The principle can be illustrated by the example of a bilayer of rigidly connected dipoles. Each layer has an isotropic tension,

upper layer: $T^u = (\gamma^u + K_B^u \alpha^u)$

Lower layer: $T^\ell = (\gamma^\ell + K_B^\ell \alpha^\ell)$ (23)

The bending moment is given by,

$$M = h^u T^u - h^\ell T^\ell \qquad (24)$$

Where h^u, h^l are the distances of the upper and lower surface from the neutral surface (which has not changed area). The total intersurface distance, h, is the sum of h^u plus h^l.

$$h = h^u + h^\ell \qquad (25)$$

In pure bending, the expansion of the upper surface relative to the compression in the lower surface is inversely proportional to the respective surface

18

compressibilities.

$$\frac{-\delta\alpha^u}{\delta\alpha^\ell} = \frac{K_B^\ell}{K_B^u} \tag{26}$$

For small curvature, the changes in the area strains $(\delta\alpha^u, \delta\alpha^\ell)$ are given by,

$$\delta\alpha^u = h^u \left(\frac{1}{R_1} + \frac{1}{R_2}\right)$$

$$\delta\alpha^\ell = -h^\ell \left(\frac{1}{R_1} + \frac{1}{R_2}\right) \tag{27}$$

R_1, R_2 are the principal radii of curvature for the surface.

Using Eqns (25), (26), and (27), the neutral surface distances are determined,

$$h^u = \frac{h}{1 + K_B^u / K_B^\ell}$$

$$h^\ell = \frac{h}{1 + K_B^\ell / K_B^\nu} \tag{26}$$

First, consider the example of the flat membrane initially in equilibrium,

$$T_0^u = 0 = \gamma_0^u + K_B^u \alpha_0^u$$

$$T_0^\ell = 0 = \gamma_0^\ell + K_B^\ell \alpha_0^\ell$$

$\alpha_0^u, \alpha_0^\ell$ are the initial fractional area contractions.
If the surface is bent, then a bending moment will be produced,

$$M = K_B^u \delta\alpha^u h^u - K_B^\ell \delta\alpha^\ell h^\ell$$

which is given by,

$$M = h^2 \left(\frac{K_B^\ell K_B^u}{K_B^u + K_B^\ell}\right) \left(\frac{1}{R_1} + \frac{1}{R_2}\right) \tag{27}$$

The bending rigidity constant, D, is

$$D = h^2 \left(\frac{K_B^\ell K_B^u}{K_B^u + K_B^\ell} \right) \tag{28}$$

Consider the second problem, an induced bending moment, created by a change in the surface chemical equilibrium (change in the interfacial free energy density). The change in surface free energy density is expressed,

$$E = \delta\gamma^u \delta\alpha^u + \delta\gamma^\ell \delta\alpha^\ell + \tfrac{1}{2}K_B^u(\delta\alpha^u)^2 + \tfrac{1}{2}K_B^\ell(\delta\alpha^\ell)^2 \tag{29}$$

This reduces to,

$$E = \left[\frac{\delta\gamma^u}{K_B^u} - \frac{\delta\gamma^\ell}{K_B^\ell} \right] \frac{D}{h}\left(\frac{1}{R_1} + \frac{1}{R_2} \right) + \frac{D}{2}\left(\frac{1}{R_1} + \frac{1}{R_2} \right)^2 \tag{30}$$

For small displacements, ζ, from the horizontal, the total curvature is given by,

$$\frac{1}{R_1} + \frac{1}{R_2} \simeq -\nabla^2\zeta \tag{31}$$

where ∇^2 is the Laplacian, differential operator ($\partial^2/\partial x^2 + \partial^2/\partial y^2$, in cartesian coordinates).

The equilibrium state is derived from the condition that the free energy is a minimum, including the work done by external tractions (e.g. normal traction or pressure on the surface).

$$\delta \int E \, ds - \int P(\delta\zeta) \, ds = 0 \tag{32}$$

where P is pressure, $\int ds$ the integral over the total surface.

Recalling

$$E = -GD\nabla^2\zeta + \tfrac{1}{2}D(\nabla^2\zeta)^2 \tag{33}$$

$$G \equiv \left[\frac{\delta\gamma^u}{K_B^u} - \frac{\delta\gamma^\ell}{K_B^\ell} \right] \frac{1}{h}$$

The variation involves two types of integrals,

$$\delta\tfrac{1}{2} \int(\nabla^2\zeta)^2 \, ds = \int(\nabla^2\zeta)(\nabla^2\delta\zeta) \, ds$$

$$\delta \int(\nabla^2\zeta) \cdot ds = \int(\nabla^2\delta\zeta) \, ds$$

The first reduces to,

$$\delta\tfrac{1}{2}\int(\nabla^2\zeta)^2\,ds = \int\delta\zeta(\nabla^4\zeta)\,ds + \oint(\nabla^2\zeta)\frac{\partial\delta\zeta}{\partial n}\,d\ell - \oint\delta\zeta\frac{\partial}{\partial n}(\nabla^2\zeta)\,d\ell \qquad (34)$$

where $\partial/\partial n$ is differentiation along the outward normal to the contour; $\oint d\ell$ is the integral around the bounding contour of the surface. The second integral becomes,

$$\delta\int(\nabla^2\zeta)\,ds = \oint\frac{\partial\delta\zeta}{\partial n}\,d\ell \qquad (35)$$

(For similar methods, see Landau and Lifshitz [5].)

Using Eqns (32), (33), (34) and (35), the surface integral components yeild the standard bending equilibrium equation,

$$D\nabla^4\zeta - P = 0 \qquad (36)$$

The contour integrals yield boundary conditions provided no constraints exist at the edge.

$$\nabla^2\zeta - G = 0 \qquad (37)$$

$$\frac{\partial}{\partial n}(\nabla^2\zeta) = 0 \qquad (38)$$

From Eqn (37), a simple consequence is observed,

$$-\left(\frac{1}{R_1} + \frac{1}{R_2}\right) = \left(\frac{\delta\gamma^u}{K_B^u} - \frac{\delta\gamma^\ell}{K_B^\ell}\right)\frac{1}{h} \qquad (38a)$$

at a free edge, which represents the balance of bending resistance against the induced bending moment (couple).

For constrained edges, the contour integrals yield the reaction shear and moment,

$$S_B = -D\frac{\partial^3\zeta}{\partial n^3}$$

$$M_B = D\frac{\partial^2\zeta}{\partial n^2} - DG \qquad (39)$$

EXAMPLES

In this section, the results of analysis for two model experiments will be given to illustrate the application of the two-dimensional material concept. Actually, the experiments were performed on erythrocyte membranes and the analyses were used to provide correlation with the observed material behaviour (details are provided in [3]).

The first example is the fluid-shear deformation of a single point attached plane membrane disk (Fig. 3). τ_0 is the fluid-shear stress (dynes/cm^2) acting on the surface; R_0 is the initial radius of the disk. Fig. 4 represents the non-uniform extension of material elements and the hyper-extension of the surface. Fig. 5 plots a dimensionless "tension" versus overall extension ratio.

The second example is the micro-pipette aspiration of a flat membrane surface of infinite extent (Fig. 6). The P's are hydrostatic liquid pressures in the regions shown. D_p is the length of the aspirated tongue; R_p is the pipette radius. Fig. 7 plots the extension ratio maximum, at the pipette tip, versus the distance the surface is aspirated; this demonstrated the simple "stretching" character of the experiment. demonstrates the simple "stretching" character of the experiment.

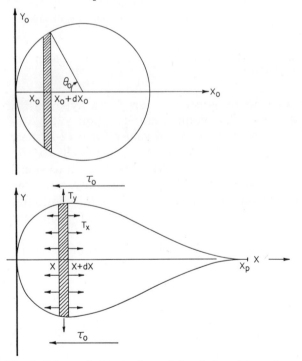

Fig. 3. Schematic illustration of the deformation of an element defined by $(x_0, x_0 + \mathrm{d}x)$ into the element $(x, x + \mathrm{d}x)$ under the action of a uniform fluid shear stress, τ_0; X_p is the point of attachment.

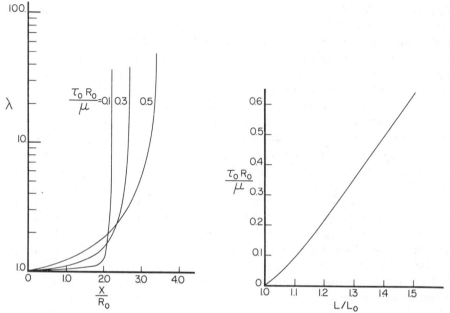

Fig. 4. The principal extension ratio λ_x is plotted against the dimensionless coordinate $\bar{\bar{x}}$ of the deformed disk for three values of the ratio $\tau_0 R_0/\mu$.

Fig. 5. The dimensionless ratio $\tau_0 R_0/\mu$ is shown as a function of the total length to original length ratio (excluding the tether length).

For both of these examples, the tension-strain law for a hyperelastic, two-dimensional material with negligible compressibility (Eqn (22)) was used. Experiments on erythrocyte membranes yield values for μ (2-D shear modulus) of $7 \cdot 10^{-3}$ dynes/cm for non-nucleated mammalian cells to $15 \cdot 10^{-2}$ dynes/cm for nucleated cells of amphibians. The two-dimensional compressibility modulus (K_B) is about 10—100 dynes/cm (only measured for mammalian erythrocytes). This data is provided for the purpose of illus-

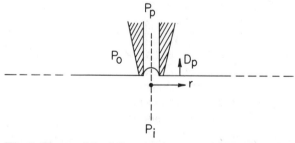

Fig. 6. The model of the membrane region surrounding the pipette tip is shown with the critical cap formed in the mouth of the pipette.

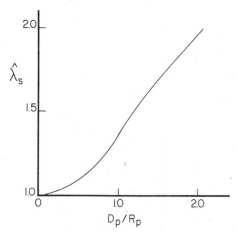

Fig. 7. The extension ratio λ_s at the pipette tip is plotted as a function of the dimensionless distance D_p/R_p.

trating the validity of the small or negligible compressibility assumption. The compressibility modulus is at least 10^3 times as large as the shear modulus; therefore, the assumption is appropriate and the simplified material relation can be used (provided that the surface area is not required to increase, e.g. under large isotropic tensions).

Comparing these elastic constants with three-dimensional material counterparts is meaningless, as previously discussed; however, a subjective relationship can be obtained by dividing the 2-D moduli by the layer thickness,

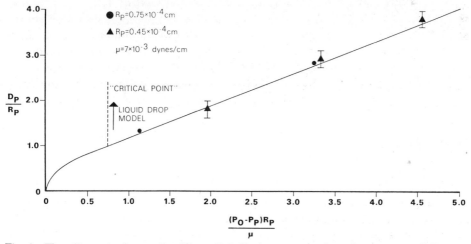

Fig. 8. The dimensionless ratio $(P_i - P_p)\, R_p/\mu$, representing the pressure difference required to aspirate the surface a distance D_p into the pipette, is shown as a function of the dimensionless distance D_p/R_p.

24

say 10^{-7} cm for a macromolucule

$\dfrac{K_B}{t}$ is approx. $10^8 - 10^9$ dynes/cm^2

$\dfrac{\mu}{t}$ is approx. $10^5 - 10^6$ dynes/cm^2

It is apparent that the effective compressibility modulus is approaching values for liquids; the effective shear modulus, on the other hand, is much lower than that of latex rubber.

Even though the experiments and applications of the material concepts have only been applied to erythrocyte membranes, the basis for other membrane systems is established. This includes the possibility of additional structural elements (e.g. microtubules, fibers, etc.); anisotropic distributions in the membrane plane can be accommodated by making μ a matrix.

The bending resistance for membrane bilayers has been developed with different properties for each layer. Using values for the two-dimensional compressibility moduli previously discussed (K_B is approx. $10-100$ dynes/cm), the negative pressure required to form a hemispherical cap, 0.5 μm in radius, would be $10-100$ dynes/cm^2. Micropipette experiments on human red blood cells yield a value of 200 dynes/cm^2 [3], indicating that the bending resistance is small compared with the shear resistance [3] in the micropipette deformation of red cells.

It was shown that the location of the neutral surface is a function of the upper and lower surface compressibility moduli. Also, these moduli may change in value when going from compression to tension. Therefore, the neutral surface will move in response to curvature reversal.

Alterations in the interfacial, chemical free energy densities can create induced moments and produce curvature. Eqn (38a) demonstrates that small changes in environment can cause appreciable curvature production because of the small layer separation. In order to produce a 1-μm radius of curvature, only an interfacial free energy density change of $10^{-2}-10^{-3}$ times the compressibility modulus K_B is required, i.e. about $10^{-2}-10^{-1}$ dyne/cm.

REFERENCES

1. Evans, E.A. (1973) A New Material Concept for the Red Cell Membrane. Biophys. J. 13, 926.
2. Evans, E.A. (1973) New Membrane Concept Applied to the Analysis of Fluid-Shear and Micropipette Deformed Red Blood Cells. Biophys. J. 13, 941.
3. Evans, E.A., and LaCelle, P.L. (1975) Intrinsic Material Properties of the Erythrocyte Membrane Indicated by Mechanical Analysis of Deformation, Blood. Vol. 45, p. 29.
4. Prager, W. (1961) Introduction to Mechanics of Continua. Ginn and Co., Boston.
5. Landau, L.D. and Lifshitz, E.M. (1970) Theory of Elasticity. Pergamon Press, London.

Comparative Physiology — Functional Aspects of Structural Materials
Eds L. Bolis, H.P. Maddrell and K. Schmidt-Nielsen
© North-Holland Publishing Company — 1975 — Amsterdam

Time dependent properties - their measurement and meaning

W.D. BIGGS

University of Reading, Reading (U.K.)

(1) INTRODUCTION

Most structural engineering is based upon the assumption that the properties of materials under stress remain constant (or, at worst, change only slightly) over a period of time. In those cases where this is not so the analysis of the structure is immensely more complicated and elaborate computer methods are often needed. With few exceptions (shell, hard coral, etc.) biological materials show time dependent behaviour — this is manifested in such phenomena as creep (progressive change of dimensions under constant stress), stress relaxation (diminution of stress at constant strain) and the anelastic damping of periodically varying stresses or strains.

Each of these phenomena has its origins in the atomic and molecular structure and there are a very large number of possible atomic mechanisms which govern them — each having its own particular time characteristic which may range from about 10^{-13} s for a single atom exchange up to 10^{-1} s for molecular chains and by a further ten orders for large groups of molecular chains.

Despite the large numbers of possible mechanisms it appears that, in biomaterials as in engineering materials, there are only a limited number of ways in which the bulk materials behave and that, in many cases, the behaviour of whole groups of materials may be represented by one fairly general equation. In these circumstances the engineer, being essentially more interested in describing the performance rather than explaining the mechanisms, finds it expedient to concentrate primarily upon the generalised continuum behaviour rather than to consider the detailed mechanisms. Faced with three variables stress, strain and time it is further expedient to consider the continuum in terms of representative models composed of elements whose behaviour can be specified as precisely as may be desired in terms, usually, of mechanical or electrical analogues.

The first part of the present paper summarises the model approach to time dependence in materials. Since it is available in the standard texts (e.g. [1,2,3]) one may ask why it is thought necessary. It has been my experience that biologists do not find the information readily accessible to them since the standard texts are either heavily mathematical or deal exclusively with man made polymers so that one of the objectives of the paper is to summarise the experimentally accessible time dependent phenomena in relatively simple model terms.

The second objective is to call for some measure of standardisation in symbology and in presentation of the data. A recent survey [4] shows that, at present, we are woefully short of hard data on time dependent biomaterials. But interest is increasing and it seems to me that an authoratative body such as this should give some general guidance and recommendations. Thus, since it is often desirable that the rheological properties of biomaterials should be compared with those of engineering polymers, the author will use (and hopes that it may encourage others to use) the symbols and nomenclature recommended by the American Society of Rheology [5]. These are substantially the same as those used by [1] and Appendix A of his book provides a definitive glossary.

(2) SOLID STATE MODELS

Basic models, capable of describing the mechanical behaviour of real solids have been postulated since the late nineteenth century and form the basis of what is now known as the theory of linear viscoelasticity.

The origins of the theory are simple enough. The basic elemental components of behaviour are taken to be ideal (Hookean) elasticity and ideal (Newtonian) viscosity. Theoretical models are then constructed out of these components and the behaviour of the resultant system is analysed. The nature of the relations between stress and strain in a spring (the elastic component) and a dashpot (the viscous component) lead to simple exponential relationships which are found to agree approximately with the experimental evidence on a wide range of materials.

By an obvious electrical analogy the relations between emf and current in a resistor and a capacitor can also be used. This method is, in some ways more convenient and more powerful but is less used, the rules for constructing equivalent electrical analogues are summarised by McCrum et al. [3], Kennedy [6], etc.

However, in adopting a model approach to the time-dependent properties of biomaterials there are three important facts to be borne in mind.

(1) For the model to be mathematically exact the material must possess certain properties:

(a) It must be incompressible; that is to say it must deform more readily in shear than in compression. This means that Poisson's ratio must be close to the upper limiting value $\nu = 0.5$ when the ratio shear modulus/bulk modulus tends to zero.

(b) It must be isotropic, i.e. the relationship between stress and strain is the same in all directions for the unstrained material. Such a material is usually (though not necessarily) also homogeneous, i.e. of uniform structure.

(c) It must be linearly viscoelastic; that is to say it must obey Boltzmann's superposition principle that stresses act independently while strains add algebraically over a given time.

It is very unlikely that any biomaterial obeys these conditions over the whole of the stress-strain range to fracture, though it appears likely that many do over particular ranges. But, frankly, we do not know and a useful research project would be the investigation of some of the 'basis' materials (silk, chitin, collagen, etc.) to see where, and how big, the divergencies are. At least this would enable us to pursue a given model whilst appreciating the magnitude of the errors. But, in addition, it would help to show the stage at which the simplistic linear assumptions become invalid and the more complex (though not necessarily more difficult) non-linear viscoelasticity must be invoked.

(2) No model is unique. Poincaré [7] has pointed out that if a physical phenomenon can be represented by one model it can also be represented by an infinite number of other models. Thus, in selecting a model, it usually pays to follow the engineer's example and select the simplest one that will represent the experimental phenomena adequately rather than precisely.

(3) It follows from (2) that we must not expect a model to be capable of interpretation in molecular terms. There are, in fact, certain cases where this is so, but these are the exception rather than the rule and, in many cases, the exercise has been misleading rather than rewarding. A model describes, but it does not generally explain and it must be accepted as a tool for describing, as precisely as we need, those features which we must, later, explain.

(3) VISCOELASTIC MODELS

Apart from the aid to visualisation there is a more profound reason for using models to describe the viscoelastic behaviour of materials. Alfrey and Doty [8] pointed out that there are seven possible ways of describing these properties and, of these, only four are general and fundamental in that they yield general differential equations which can be solved for a wide variety of transient conditions. Here we consider briefly three methods; these three only are related formally to the methods which are available to the experimenter. They are (a) the Voigt model (b) the Maxwell model (c) the impedance function.

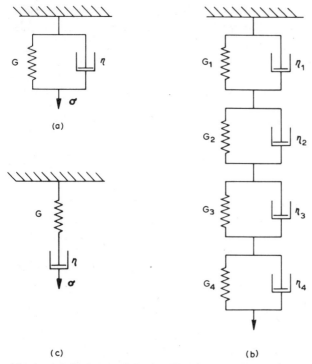

Fig. 1. (a) Voigt model. (b) Maxwell model. (c) Non-degenerate four-element Voigt model.

(a) The Voigt Model

In its simplest form this consists of an elastic spring and a viscous dashpot joined in parallel (Fig. 1a). It is the simplest model which will represent retarded elasticity and creep and it is completely specified by two parameters: the modulus of the spring G and the viscosity of the dasphot η. The strain γ associated with a stress σ is given by the differential equation

$$\sigma = G\gamma + \eta\dot{\gamma} \tag{1}$$

Clearly if $\eta = 0$ the model reduces to an elastic element while if $G = 0$ it reduces to a viscous element; such models are called "degenerate". The problem of creep can be formulated as the relationship between $G(t)$ and $\gamma(t)$ under a constant stress σ_0. Thus we write Eqn (1) as

$$\sigma_0 = G(t)\,\gamma(t) + \eta\dot{\gamma}(t)$$

whence

$$\frac{\sigma_0}{\eta} = \frac{\gamma(t)}{\tau} + \gamma(t) \tag{2}$$

where $\tau(=\eta/G)$ is the retardation time.

This is a linear differential equation which may be made exact and then integrated between limits $\gamma(0) = 0$ and $\gamma(t) = t$ using the integration factor $\exp(-t/\tau)$. This gives

$$\gamma(t) \exp(-t/\tau) = \frac{\sigma_0}{G} \, [\exp(-t/\tau) - 1] \tag{3}$$

whence

$$\frac{\gamma(t)}{\sigma_0} = J(t) = J[1 - \exp(-t/\tau)] \tag{4}$$

where $J(t) = 1/G(t)$ is the "creep compliance". More realistic models are obtained when several Voigt elements are joined in series. Specification of the four non-degenerate elements shown in Fig. 1(b) now requires a knowledge of eight parameters $G_1\eta_1$; $G_2\eta_2$; $G_3\eta_3$; $G_4\eta_4$: these each have a constitutive equation of the same form as Eqn (1). In general terms

$$\sigma = G_i\gamma_i + \eta_i\dot{\gamma}_i$$

and, for n such elements,

$$J(t) = \sum_{i=1}^{n} J_i[1 - \exp(-t/\tau_i)] \tag{5}$$

In the limit the discrete set of Voigt elements may be replaced by a continuous distribution of elastic compliance as a function of retardation time. That is to say the finite set of physical constants $G_1\eta_1$; $G_2\eta_2 \ldots G_n\eta_n$ may be replaced by a continuous function $J(\tau)d\tau$ which tells how much of the elastic compliance has retardation time in the range $(\tau + d\tau)$. Thus

$$J(t) = \int_0^\infty J(\tau) \exp(-t/\tau) \, d\tau \tag{6}$$

to which the appropriate expressions for one or two degenerate elements may be added.

Experience has shown that a logarithmic scale is more convenient. Accordingly the continuous distribution of retardation time is defined as $L(\ln\tau)$, i.e.

the compliance is described in terms of the contribution made by those processes having retardation times lying between $\ln \tau$ and $\ln \tau + d(\ln \tau)$. Eqn (6) is now written as

$$J(t) = \int_{\ln \tau = -\infty}^{\ln \tau = \infty} L(\tau)[1 - \exp(-t/\tau)] \, d \ln \tau \tag{7}$$

which may be amended, as before, to account for up to two degenerate elements. The distribution function $L(\tau)$ (often called the creep function or the retardation spectrum) has dimensions of compliance. It can be made dimensionless though, as Ferry [1] points out the added mathematical convenience (the integral over $d \ln \tau$ from $-\infty$ to ∞ is unity) is more than outweighed by the difficulty of applying the dimensionless function to experimental data and recommends that the non-normalised function $L(\tau)$ be used. Some authors have used $d \log_{10} \tau$ rather than $d\ln \tau$ so that the numerical results differ by a factor of 2.303. Provided the base is made clear this is no more than a petty inconvenience but, in the interests of uniformity, one that is better avoided.

(b) The Maxwell Model

This consists of a spring and dashpot joined in series (Fig. 1(c)), it is the simplest model which will represent stress relaxation. The equation of motion is given by

$$\dot{\gamma} = \frac{\dot{\sigma}}{G} + \frac{\sigma}{\eta} \tag{8}$$

and since, in a stress relaxation experiment $\gamma = 0$ after the initial strain γ_0 has been applied we obtain

$$\frac{\sigma(t)}{\gamma_0} = G(t) = \frac{\sigma}{\gamma_0} \exp(-t/\tau)$$

i.e.

$$G(t) = G \exp(-t/\tau) \tag{9}$$

Thus, by analogy with Eqn (5) we obtain for n non-degenerate elements

$$G(t) = \sum_{i=1}^{n} G_i \exp(-t/\tau_i) \tag{10}$$

As before we consider a continuous distribution $G(\tau)$ dτ describing the amount of elastic modulus which has relaxation time in the range τ + dτ and, for convenience, define a relaxation function $H(\tau)$ to obtain an expression of similar form to Eqn (7)

$$G(t) = \int_{\ln \tau = -\infty}^{\ln \tau = \infty} H(\tau) \exp(-t/\tau)\, \mathrm{d} \ln \tau \tag{11}$$

The general remarks concerning the creep function $L(\tau)$ also apply to the relaxation function $H(\tau)$.

(c) The Impedance Function

The relationships described above have an immediate electrical analogy. If we assume that stress corresponds to voltage, viscosity to resistance, strain rate to current and $-G/\omega$ to the reactance, the steady-state response of the mechanical model to a sinusoidal stress $\sigma = \sigma_0 \cos \omega t$ can be related to a complex impedance $Z = a + ib$ where a is the "viscous resistance" (corresponding to η) and b is the "elastic reactance" (corresponding to $-G/\omega$). Both a and b depend upon frequency. The complex 'impedance function' $Z(\omega)$ defines the properties of a viscoelastic material in a manner which is completely equivalent to those of both the Voigt and the Maxwell models and although its most obvious function is in connection with the steady-state response to sinusoidal stress the use of Fourier and Laplace transforms makes it as general and fundamental as both models in its definition of the viscoelastic properties.

(4) MODEL BEHAVIOUR AND EXPERIMENTAL METHODS

Determination of the long time properties of viscoelastic materials is generally done by stress relaxation in which the decay of stress at constant extension is measured or by creep involving the increase in strain at constant stress. The theory indicates that in both tests the initial condition should be applied instantaneously, this is difficult to achieve but, providing that the time response is long (relaxation times of at least $\tau = 10^3$ s) the finite time taken to apply the initial stress is usually unimportant. For materials with shorter relaxation times the method of reduced variables (the so-called "shift function") as described below can be used if its application can be properly validated.

We consider here only stress relaxation, though the same description applies to creep. A single Maxwell model having $G = 10^8$ Nm^{-2} and $\eta = 10^8$ Nsm^{-2} has a relaxation time $\tau = 1$ s and yields the relaxation curve

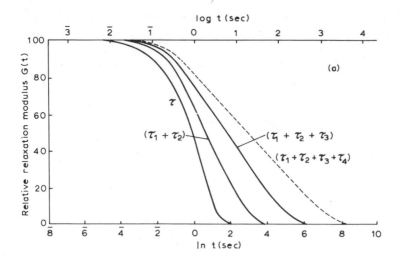

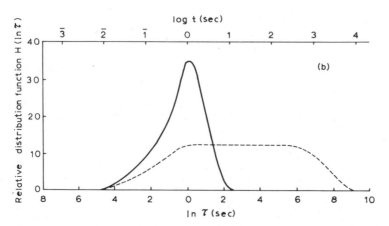

Fig. 2. (a) Family of stress relaxation curves for 1-, 2-, 3- and 4-element Maxwell models having relaxation times of 1, 10, 100 and 1000 s. (b) Distribution functions for one and four element models shown in (a).

shown by the solid line in Fig. 2(a). The distribution function $H(\tau)$ should, ideally, be computed using Eqn (11) but a simpler graphical method [8] is often adequate providing there are no sharp discontinuities in the relaxation curve. This method simply involves plotting the gradient of the relaxation curve against $\log \tau$ or, preferably $\ln \tau$. The distribution function for the single model is then as shown in Fig. 2(b). The curve is approximately log-normal and shows that the relaxation modulus is dominated by a process having relaxation time $\tau = 1$ s, as, indeed, it should.

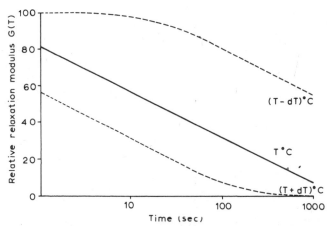

Fig. 3. Stress relaxation of four element model of Fig. 2(a) as 'seen' experimentally (solid curve). The shift function enables the remainder of the curve to be obtained within experimentally accessible times.

Consider now the progressive addition of further elements; for simplicity we assume that the spring constant of each is the same as before ($G = 10^8 Nm^{-2}$) but that the viscosities differ so that the additional elements have relaxation times of 10, 100 and 1000 s. The family of relaxation curves for the four models is shown (plotted to the same relative scale of $G(t)$) in Fig. 2. The distribution curve for the four element model is flattened and shows equal contributions from processes having relaxation times in the range $1-10^3$ s. Clearly, with n elements a long, flattened distribution over many orders of time will arise. This is well shown, for instance in the relaxation function for the cell wall of *Nitella opaca* [9].

Experimental methods for determining the relaxation time are fully discussed by e.g. Ferry [1], McCrum et al. [3] and will not be pursued here. One important experimental technique is, however, worth noting. It can be seen from Fig. 2(a) that determination of the full spectrum of the four-element model would require experimental observations over a time scale from 10^{-2} to 10^4 s whereas the experimentally convenient time scale for such tests is, perhaps $1-10^3$ s. If such a restriction were to be placed the relaxation curve for this model would appear as shown in Fig. 3 (solid curve). But we may assume that a change in temperature would be likely to have a much greater effect upon the viscosity of the dashpot than upon the modulus of the spring since, in general, the viscosity of a Newtonian fluid varies with temperature according to a function of the type

$$\eta \propto \exp \left(\frac{\Delta H}{RT} \right)$$

where ΔH is an activation energy, T is temperature and R is the universal gas constant. Thus an increase in temperature causes a uniform change in the viscosity of each dashpot (this corresponds to a uniform "shift" in the relaxation time of each element and the whole curve is shifted relative to the time axis). By this means the longer and shorter relaxation processes can be brought within the experimentally convenient time scale to yield a group of experimental curves as shown in Fig. 3 (dashed curves). These may then be combined to yield a so-called "master curve". In real solids certain complications arise (e.g. the need for a vertical shift due, usually, to a change in the value of the modulus especially in materials whose modulus is controlled primarily by the configurational entropy) and some caution is needed in interpreting the results. Full discussion of the shift method (method of reduced variables) will be found in the standard texts.

Use of the shift function has not, however, been uniformly successful in evaluating the properties of biomaterials though it has worked quite well in some of the simpler cases such as the cell wall of *Nitella opaca* [9]. There are probably several reasons. One is that the range of temperature which may be used for biomaterials is considerably narrower than for polymers so that the differences in slope of the experimental curves are small and errors are easily introduced. Another is that the initial thermodynamic state of differing samples may not be the same unless some preconditioning is applied — we shall discuss this further in Section (6) below. But probably the most important reason is that most biomaterials are fairly complex mixtures and, even in the case of polymers, the method of reducing variables in this way is not always successful. The reason is not hard to postulate. Each component of the mixture has its own spectrum of relaxation times and hence of activation energies and the overall behaviour is determined by the slowest process. But the extent of the contribution made by the slowest process is further determined by the amount of that component present. There is comparatively little information available here except in the case of some simple polymeric materials such as mixtures of polysulphide rubbers and more comprehensive data are needed if analogies are to be drawn. Finally, and superimposed upon this, is the fact that the physical distribution of the component phases varies widely (from more or less randomly particulate to more or less randomly fibrous) the first may approximate to a filled polymer, the second to a fibre composite. Our knowledge (both theoretical and experimental) of the mechanical behaviour of such mixtures is largely limited to the simpler cases where one of the phases is substantially Hookean (at least over its working range). The problems of non-linear elastic reinforcement in a viscoelastic matrix have scarcely been touched and, since most biomaterials are of this type, much remains to be done.

Whilst the discussion above refers specifically to stress relaxation the arguments and methods apply equally to creep and we shall not discuss this separately here. The relaxation time(s) can be determined directly by substituting the experimental data into the appropriate model (this will be discussed in Section (5) below.

But stress relaxation and creep represent particular solutions to be generalised viscoelastic equations and, in order to describe the material fully, we need some data which are general and fundamental. These are provided by the impedance function which must always be regarded as the primary testing method, not only because it discloses fundamental relationships in time-dependance but also because it is the only convenient method which, without recourse to shift functions, can be used to monitor both the "fast" and "slow" responses. The techniques, and the significance of the results are fully described in the literature and will not be discussed here. We shall merely point out that, under sinusoidal forcing, it offers a means by which the in-phase (storage) modulus G' may be separated from the out-of-phase (loss) modulus G''. The former corresponds to the strictly elastic part of the model, the latter to the viscosity-modified part. The two are related via the familiar equations

$$\sigma = \gamma_0 \, G' \sin \omega t + \gamma_0 \, G'' \cos \omega t$$

for a forcing function $\gamma = \gamma_0 \sin \omega t$ where γ_0 is the maximum strain in each cycle and ω is the circular frequency. Using the obvious electrical analogy the complex modulus

$$G^* = G' + iG''$$

Thus the equations of motion for the Voigt and Maxwell models may be approached experimentally for, knowing G' and G'' from dynamic tests and τ from either stress relaxation or creep tests η may be determined and the appropriate model found.

But dynamic testing offers one further and important piece of information — it is the most convenient way of 'mapping' the properties over a wide time spectrum and it yields information about particular relaxation processes. Thus, in dynamic tests on mesogloea (a collagenous connective tissue) of the sea anemone *Metridium senile* Gosline [10] has found damping peaks that correspond with particular forcing frequencies so that each peak must reflect a time-dependent process which is occurring within that same frequency range. Much more work is needed here if these processes are to be identified and there can be little doubt that dynamic testing and the consequent establishment of the impedance functions offers a promising, and greatly needed, area for research.

(5) WHICH MODEL TO CHOOSE

This depends partly on what we are looking for and also partly upon

36

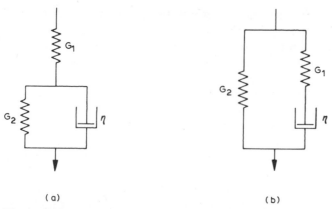

(a)   (b)

Fig. 4. Simples form of model which will demonstrate stress relaxation and creep. The two models are equivalent.

experimental convenience. Although the mathematical relationships show that formal relationships exist between the Maxwell and Voigt models the present author subscribes to the view expressed by Alfrey and Doty [8] that to use these is time consuming, inconvenient and, sometimes misleading. The simple Maxwell model cannot be used to represent creep, the simple Voigt model cannot be used for stress relaxation. But so long as our objective is simply to describe the behaviour it is rarely necessary to go further than adding one degenerate element to either — these three element models (Fig. 4) are equivalent and can represent all the observed modes of behaviour — elastic, creep, relaxation and recovery. The general equation of the model shown in Fig. 4 (the so called "standard linear solid") is

$$\sigma + \tau\dot{\sigma} = G_2 \left[\gamma + \eta \left(\frac{G_1 + G_2}{G_1 G_2} \right) \dot{\gamma} \right]$$

and, by inserting the appropriate boundary conditions (e.g. $\sigma = \sigma_0$ = constant for creep, $\dot{\gamma} = 0$ for stress relaxation) Eqn (13) can be solved for a wide variety of conditions. Furthermore it will be noted that this is the simplest model which can be used to represent the two extreme equilibrium states: that corresponding to the "relaxed modulus" G_2 (when $\dot{\gamma} \to 0$) and the "unrelaxed" or "instantaneous modulus" $(G_1 + G_2)/G_1 G_2$ (when $\dot{\gamma} \to \infty$).

But it is obvious that even a three element model cannot properly represent a material having a spectrum of relaxation times. It is, of course, perfectly easy to add more and more terms and there are many standard methods available (see, for example [11—13]). But how closely do we need to evaluate the various components of the relaxation or creep function? Clearly if, by evaluating all the different terms we can positively identify the molecular processes which are involved, the exercise is worth while but, in the present stage of knowledge this is rarely possible. Indeed, as mentioned earlier, the immense number of microscopic possibilities seem to be irrelevant because the number of macroscopic demonstration is small. We may too easily fall into the habit of straining at real gnats while swallowing hypothetical camels.

Alexander's papers on mesogloea [14] typically provides a salutory illustration of the need for economy in analysing data. His creep curves showed that the retardation times for the mesogloea of *Metridium* and *Calliactis* differed by an order of magnitude but that a very simple 'shift' (in this case normalising the function t/τ) superimposed the experimental data. The fit, however, is poor at regions of maximum curvature and I have reanalysed his data to find that the points are better fitted by (at very least) a five-element, singly degenerate model with retardation times in the range $\tau = 1-10^4$ s. But, on plotting the distribution function, I find that the compliance is dominated by processes having a retardation time $\tau = 10^{3.5}$ s whereas Alexander's simpler model, based upon the mean of experimental observations, suggested $\tau = 10^{3.4}$ s. Thus the additional work and the selection of a more complex model has told me no more about molecular interactions, fibre slippage, water content or whatever and it does not affect, in any significant detail, Alexander's conclusions based upon the simpler model. It is essential to wield Occam's razor directly and ruthlessly in setting up model representations. Indeed this may well be true of the whole area of biomechanics and of biology, the simplest model that is just adequate is the best. At least it is likely to be the most comprehensible.

But all this raises a further question namely whether the Voigt or the Maxwell model may be considered as superior from the point of view of an interpretation of behaviour based upon molecular theory. Alfrey and Doty considered this very carefully and concluded that there was no prima facie reason for preferring one to the other but that their subjective judgment came down on the side of the Voigt model. Their reasons were, essentially, that it is easier to visualise and identify flow processes in terms of configuration changes (i.e. changes in dimension, i.e. strain) in response to an applied constant stress than in terms of various components 'balancing' the stresses resulting from the imposition of a given strain. It is difficult to argue against this though there is a growing mass of experimental evidence on other materials suggesting that strain (or more precisely the effect of cumulative increments of strain) is more significant than stress, i.e. that the behaviour of many materials is dictated by the attainment of a critical, cumulative strain rather than by the need to support an externally applied stress. However this is an argument which falls outside of the present paper and will be presented at another time.

But a preference for the Voigt model goes rather further than this. In practical experimental terms it is not easy to visualise stress relaxation in shear whereas shear creep presents few problems. And there is no doubt in the author's mind that, wherever and whenever possible, tests in shear are infinitely preferable to tests in tension. For one thing shear is a homogeneous state of stress involving changes of shape without change in dimension. But, overriding these is the fact that the tensile modulus is dictated almost entirely by the shearing characteristics of the material so that, in using tension as a testing mode, we are simply studying shear at second hand.

(6) TENSION TEST

We have not, so far, discussed tension testing specifically. Despite the earlier strictures about tension the fact remains that, as an experimental method, it is simple and requires no elaborate apparatus (this makes it especially attractive for field tests). But the very nature of the test requires that the loading be continuous so that relaxation phenomena are continuing during the test itself; that is to say that every fresh increment of stress is applied to a material which is in a different thermodynamic state from the one which existed prior to the application of the previous stress increment. Thus the specimen can only be considered as being in equilibrium under very fast or very slow rates of loading, not only are these difficult to achieve experimentally but it is clear that, in the limit, they correspond to the creep and stress relaxation methods described above. It is, of course, possible to convert tension data into terms of equivalent creep and relaxation data (e.g. [1]) but numerous assumptions, not always justifiable, must be made.

Nonetheless tension tests have a real place in the overall testing strategy, partly as quick and easy sorting tests but, more importantly, as a means of providing essential information on the starting state for creep, stress relaxation or dynamic testing.

Successive loading-unloading-reloading cycles on a wide range of biomaterials by the present author and others have shown that, after a number of cycles, the material settles down to a substantially reproducible behaviour which may, in fact, be quite different from that displayed on the first, in vitro, test (Bergel, 1972, Vincent, 1974, etc.). Many workers have used tension cycling as a form of "preconditioning" a specimen prior to tests in vitro and opinions as to the validity of this prior treatment differ widely. Some of the suspicion arises from the warrantable assumption that the material, after preconditioning is not in the same condition as when removed from the organism and certianly the appearance of wide variations in preconditioning methods lends support to this view.

I do not see this as a valid objection. It seems to me that, in vivo, the material must be in thermodynamic equilibrium with its surroundings, were this not so it must spontaneously change its thermodynamic state until it is. Thus it would seem that the behaviour of any element of material in vivo is determined by very small changes about an equilibrium situation so that, on testing in vitro, we must endeavour to achieve, so far as is possible, an equilibrium state upon which to base our assessment of behaviour. It won't, of course, be the same equilibrium state, but we are concerned with changes and not with absolute values. Indeed we have no absolute values from which to work so that we can do no more, in vitro, than try to achieve a "most probable" (i.e. highest entropy) starting state since this probably mirrors (without necessarily precisely reproducing) the in vivo condition. It is here, more than anywhere else, that tensile tests (because of their convenience) have a significant part to play in the overall strategy.

These considerations lead to the further thought that, the additional factors of chemical and/or electrical balance in vivo may suggest that biomaterials can be described by a mechanical equation of state analagous to that for gases. This is a concept which has been a pipe dream of metallurgists and materials scientists for many years. Experiments have shown that the instantaneous state of such materials may be described in terms of the stress, strain, temperature and strain rate by an expression of the form

$$f(\sigma, \epsilon, \dot{\epsilon}, T) = 0 \tag{15}$$

However, the concept, where it applies at all, does so only over a very limited range of conditions in crystalline solids but over a rather wider range in polymers as, in fact, is implicit in the Boltzmann superposition principle. If it were possible to show that a biomaterial could be described, even approximately, in terms such as those of Eqn (15) it would provide a very firm basis upon which to assess the differences which might be expected between behaviour in vivo and tests in vitro and the results of tests could be extrapolated to the plant or animal with more certainty than is possible at present.

(7) DISCUSSION

After all of the foregoing it is, once again, pertinent to ask why the model approach is advocated as a means of tackling the problems of biomaterials and of biomechanics. One answer has already been given, the Maxwell and Voigt models, despite their apparent simplicity, provide two of the four methods of describing time-dependent properties in general and fundamental terms which can be applied (given patience and possibly a computer) to any set of transient conditions. But there is another reason which is, perhaps, more germane to the objectives of the present meeting. It is, essentially, that of trying to find an interface between those activities which explain but do not describe and those which describe but do not explain.

Classically speaking biology is a descriptive subject and it is interesting to recall that when the present author started to study metallurgy (before it was dignified by the now fashionable term 'materials science') much the same situation existed. Comparatively few people were working at the atomic level and, in any event, their theories were only acceptable if they enabled you to do things better, i.e. if a more specific understanding of atomic mechanisms helped you to achieve a desired result by design ab initio rather than by piecewise development.

The real breakthrough occurred about 30 years ago when the atomic theory of crystalline solids permitted — nay, encouraged — the production of many research theses dealing, in every increasing detail, with the atomic mechanisms. And each represented a diminishing return. As an engineer I want to design a whole bridge, a whole structure, a whole aeroplane. I am interested, primarily, in how the whole system behaves — and atomic or molecular theory is only useful to me if it helps to predict the overall behaviour. And, frankly, it only does so in a limited way. After some 30 years of research at the atomistic level the gap between the "whole structure" man and the "atomistic structure" man is wider than before.

You must forgive me the digression, but I make it because I feel, in my ignorance, that the study of biomaterials and biomechanics is now in the same position. It is gratifying and exciting that it should have moved so thoroughly from classification into measurement and analysis. However, analysis without explanation is sterile just as description without comprehension is empirical and we may well be so obsessed with the analysis of the trees that we are in danger of forgetting the nature of the forest that they stand in. The ultimate requirement is that of understanding the macroscopic behaviour of the plant or the animal, not the understanding of the specific characteristics of the piece, fascinating though this may be.

The challenge here is considerable. The plant, or animal exists, and, presumably, it exists successfully so that the whole system must have achieved some sort of equilibrium with its surroundings. But this does not necessarily mean that every element is in equilibrium at all times, indeed the concepts of modern physics suggest that, if certain statistical variations did not exist, some important phenomena could not, in fact, occur.

I am deeply conscious of the ultimate desirability of explanation, at whatever level is possible. But I seriously question the need. In another paper my colleague, Jim Gordon, suggests that (a) certain limitations are set by the very nature of the structure of biomaterials and, more importantly, that (b) because of these limitations certain effects, in particular an instability under internal pressure, follow inexorably. Now I suggest that this is of overriding importance. The plant or the animal is stuck with certain basic building blocks which must lead to certain types of stress-strain relationship. And these have certain consequences which can be analysed and discussed. I feel that, at the present time, the important thing is not to wonder why these characteristics exist (for even if you knew you still could not change them), but to wonder what are the inevitable consequences of having to live with these conditions.

My interest in using models stems entirely from the fact that I am much more interested in "how" rather than "why". The whole animal excites me because, ultimately, this is the object that must live with, combat, and adapt to, the environment. Stress, strain and time are, perhaps, only a part of the environment but I would suggest that if you eliminate these there is not very much left for the organism to resist.

And, because stress, strain and time are macroscopic and continuum conceptions and because most plants and animals live in a time continuum, I suggest that the macroscopic model approach may well offer a broader, if less specific, approach and one that is, in many ways, closer to the real behaviour of the organism. At very least it offers a method by which continuum behaviour may be quantitatively described, and, in so doing, it should point more clearly to those atomistic processes which need to be investigated if description is to be backed by explanation.

(8) REFERENCES

1. Ferry, J.D. (1970) Viscoelastic Properties of Polymers, 2nd Edn., Wiley, New York.
2. Aklonis, J.J., MacKnight, W.J. and Shen, M. (1972) Introduction to Polymer Viscoelasticity. Wiley/Interscience, New York.
3. McCrum, N.G., Read, B.E. and Williams, G. (1967) Anelastic and Dielectric Effects in Polymeric Solids. Wiley, New York.
4. Wainwright, S.A., Biggs, W.D., Currey, J.D. and Gosline, J.M. (1975) Mechanical Design in Organisms. Edward Arnold, London, in the press.
5. Leaderman, H. (1957) Trans. Soc. Rheol. 1, 213.
6. Kennedy, A.J. (1962) Processes of Creep and Fatigue in Metals. Oliver and Boyd, London.
7. Poincaré, H. (1929) The Foundations of Science. Science, 181.
8. Alfrey, T. and Doty, P. (1945) Methods of Specifying the Properties of Viscoelastic Materials, J. Appl. Phys. 11, 700.
9. Haughton, P.M., Sellen, D.B. and Preston, R.D. (1968) Dynamic Mechanical Properties of the Cell Wall of *Nitella Opaca*. J. Exp. Bot. 19, 1.
10. Gosline, J.M. (1971) Mechanics of Mesoglea in the Sea Anemone, *Metridium senile*. J. Exp. Biol. 55, 775.
11. Lipka, J.L. (1918) Graphical and Mechanical Computation. Wiley, New York.
12. Whitehead, J.B. (1935) Impregnated Paper Insulation, Wiley, New York.
13. Morrow, C.T. (1969) Non-linear Transient and Dynamic Behaviour in Bovine Muscle. Ph. D. Thesis, Pennsylvania State University.
14. Alexander, R.McN. (1964) Viscoelastic Properties of the Mesogloea of Jellyfish. J. Exp. Biol. 41, 363.

Comparative Physiology — Functional Aspects of Structural Materials
Eds L. Bolis, H.P. Maddrell and K. Schmidt-Nielsen
© North-Holland Publishing Company — 1975 — Amsterdam

Structural proteins in relation to mechanical function: Crosslinks in insect cuticle

SVEND OLAV ANDERSEN

Zoophysiological Laboratory C, August Krogh Institute, University of Copenhagen (Denmark)

Most extracellular structural proteins are stabilized by the presence of covalent cross-links between the peptide chains, and the cross-links are formed after the proteins have been secreted and organized in the extracellular space. The functional importance of the cross-links may vary, but in general they make the proteins more insoluble, more stiff and hard, and more resistant to enzymatic degradation. A variety of chemical structures have been reported which function as natural cross-links, and one of the few things they have in common is that they are formed outside the cell. This poses the question: if the degree of cross-linking is an important factor in determining the mechanical properties of the extracellular materials how can the cells know how many cross-links to introduce and how can they control the degree of cross-linking? is there some form of feed-back mechanism from the extracellular material to the cells regulating the synthesis of cross-links? We have for some years been investigating the insect cuticle, a system where I believe that some form of control is present, regulating both the rate of cross-linking, the degree of cross-linking, and may be even the type of cross-links formed. In insects there are pronounced regional differences in the degree of cuticular cross-linking, and these differences have apparently a functional significance. The control mechanisms have not yet been unraveled, but feed-back phenomena appear not to be involved. The cuticle is an extracellular structure formed by a single-layered epidermis; it consists mainly of protein and chitin plus some lipid, the protein content ranges from about 50 to 80%. The structure, composition, and properties of the cuticle vary to some extent according to what part of the body it comes from. Table I gives some of the characteristic properties of different types of cuticle found in locusts. Similar types are present in other insects, but my personal experience is largely based upon a study of locust cuticle, and the discussion will therefore mainly be concerned with that animal.

TABLE I

VARIOUS TYPES OF CUTICLE PRESENT IN THE LOCUST, *S. GREGARIA*

Cuticle	Properties
Main type of cuticle from head, thorax, legs, wing and abdomen	Hard, colourless, and opaque.
Mandibles, tibial spines	Hard, brown, and brittle.
Melanin-containing areas on wings and femur	Hard, and black.
Arthrodial membranes	Flexible, inextensible, and colourless.
Abdominal intersegmental membranes in sexually mature females	Flexible, extensible, and colourless
Cornea from compound eye	Solid and transparent.
Resilin-containing ligaments.	Long-range elastic and transparent.

The properties of the various types of cuticle are determined by a number of factors: protein and chitin content, organisation of the protein and chitin molecules, properties of the individual proteins, types of cross-links, number of cross-links, and the presence of other molecules. What makes insect solid cuticle a nearly ideal system for a study of the importance of cross-links is that it is possible to obtain a given cuticular structure with different degrees of cross-linking. The cuticle is deposited in an un-cross-linked state during the last few days before the insect moults, and the formation of cross-links starts soon after emergence and continues for a shorter or longer period. Femur cuticle from adult locusts increases its degree of sclerotization for several weeks, and the material introduced as cross-links eventually makes up about 3% of the dry weight of the cuticle. It is thus possible to correlate the properties of the cuticle with the degree of sclerotization. But before I go further into this it may be better to say something about the nature of the cross-links and how they are formed.

The immediate precursor of the cross-links is *N*-acetyldopamine, which appears to be formed in the haemocytes and transported into the cuticle via the epidermal cells. In the cuticle the compound is activated at its β-carbon atom by means of an enzyme already present in the cuticle [1], and the activated compound reacts with free amino and phenolic groups to give cross-links of the general structure:

Acid hydrolysis will split these cross-links with the simultaneous release of various ketocatechols, such as:

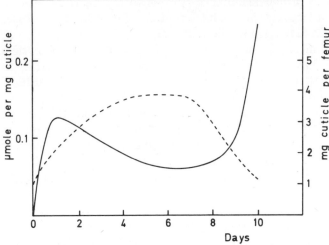

When I talk about degrees of sclerotization I refer to the amounts of such ketocatechols which we have obtained on acid hydrolysis. Other types of cross-links, which may occur in the cuticle, will therefore not be taken into account in these determinations.

We have determined the degree of sclerotization both of a single type of cuticle at various times during sclerotization and of various types of cuticle at a given moment during maturation [2—4]. Figs. 1 and 2 show how the weight and the degree of sclerotization change with time in femur cuticle from 5th instar larvae and from adult locusts. Just after emergence no keto-catechols can be obtained but soon they appear and increase rapidly in amount. In adult cuticle they go on increasing for a long period, in larvae only for a single day. This may be important for the future fate of the cuticles. A major part of the larval cuticle will be digested later and the

Fig. 1. Changes in weight and ketocatechols in femur cuticle from 5th instar locusts in relation to the age of the animals. Fully drawn line: amounts of ketocatechols liberated from cuticle by hydrolysis in 1 M HCl for 3 h. Broken line: dry wt of cuticle from one femur. Day 0: moult from 4th to 5th instar. The animals moulted to adults during the 10th day. (Redrawn from Andersen [2].)

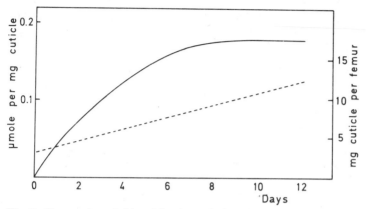

Fig. 2. Changes in weight and ketocatechols in femur cuticle of adult locusts in relation to the age of the animals. Fully drawn line: amounts of ketocatechols liberated from cuticle by hydrolysis in 1 M Hcl for 3 h. Broken line: dry weight of cuticle from one femur. Day 0: moult from 5th instar to adult. (Redrawn from Andersen and Barrett [4].)

insoluble part discarded at the next moult, whereas the adults stay with their cuticle for the rest of their lives. They can therefore afford to stabilize it thoroughly.

I have made no measurements of the mechanical properties of femur cuticle myself, but Hepburn and Joffe [5] have determined the stiffness and breaking strength of locust femur cuticle and have compared their results with my determinations of the degree of sclerotization. Their measurements show that the mechanical properties change parallel to the decrease in solubility of the proteins, indicating that the introduction of cross-links increases both stiffness and breaking strength of the material.

When it comes to comparing the properties of various types of cuticle from a single animal I have no measurements of mechanical properties to refer to, and I must restrict myself to a subjective estimate of hardness obtained by handling the various types of cuticle. Table II gives the degree of sclerotization of various cuticles both from locusts and from a beetle species. The table shows that all types of cuticle investigated are sclerotized, that the degree of sclerotization varies considerably, and that the types of cuticle where strength is of functional importance are also the most heavily sclerotized samples: the mandibles and parts of the thorax involved in the flight system.

The different degrees of sclerotization in the various types of cuticle could be caused by different concentrations of cross-linking enzyme. Measurements of enzyme activity in the cuticle show, however, that there is no relationship between enzyme content and degree of sclerotization or rate of sclerotization. It could also be the amount of substrate, N-acetyldopamine, which is the rate limiting factor. To test the idea we injected both young and older locusts with labelled precursors for the cross-links and determined the

TABLE II

AMOUNTS OF KETOCATECHOLS OBTAINED FROM CUTICLE LOCUSTS, *S. GRE-GARIA*, AND OF ROSE CHAFERS, *PACHYNODA EPPHIPIATA*

The results are expressed in percentage of the dry weight of the samples. n.d., not determined.

	Locust, 5th instar larva	Locust, adult 10—12 days	Rose chafer
Tibia	1.71	2.41	n.d.
Femur	0.99	2.65	3.82
Mandibles	3.74	3.36	n.d.
Pronotum	0.39	2.46	4.05
Dorsal thorax	0.75	5.25	4.40
Ventral thorax	0.37	1.51	4.48
Thoracic ribs	n.d.	4.47	n.d.
Dorsal abdomen	0.24	0.33	0.98
Ventral abdomen	0.49	0.27	2.69
Forewings	1.29	1.97	3.38
Hindwings	1.21	2.02	0.82
Cornea	n.d.	0.22	n.d.
Intersegmental membranes	n.d.	0.20	n.d.

degree to which they became incorporated into the various types of cuticle [3]. A good agreement was found between the amounts of radioactive dopamine incorporated into the cuticle of young adult locusts and the total amounts of cross-links present in corresponding samples of cuticle from untreated animals of the same age. This relationship is no longer valid in older animals, indicating that the incorporation rate changes independently in the various types of cuticle during maturation.

My conclusions are that both the initial rate of sclerotization and the changes with time in the rate of sclerotization are governed by local factors which presumably reside in the epidermal cells. The timing of the start of sclerotization, however, appears to be determined centrally by the hormone, bursicon, secreted from the central nervous system soon after emergence. Although a feed-back mechanism from the cuticle to the epidermis may exist we have so far not found any evidence for its presence, and I am in favour of the hypothesis that all the necessary and detailed information about what and how much shall be done during cuticle formation is present in the epidermal cells.

In this discussion I have so far avoided the complications arising from our recent findings that at least two different types of cross-links can be formed simultaneously in the same cuticle and from the same substrate. One is the type discussed here where the sidechain of *N*-acetyldopamine is used to establish cross-links between peptide chains. We have now obtained evidence that cross-links can also be formed where the ring of *N*-acetyldopamine is involved [6]. This cross-linking mechanism corresponds to the classical quinone tanning of insect cuticle [7]. We are not yet able to determine the quantitative importance of this system for the hardening of insect cuticle. We are, however, able to measure the activities of the two enzymes involved in sclerotization, and our results indicate that sidechain-tanning dominates in colourless or lightly coloured cuticles, and that quinone-tanning dominates in dark brown cuticles. It is my impression that cuticle where quinone-tanning dominates tends to be more brittle than cuticle which has been sclerotized mainly by the sidechain pathway. However, these results are still preliminary and they will have to be supplemented by investigations on more insect species before any firm conclusions can be drawn.

REFERENCES

1. Andersen, S.O. (1972) An enzyme from locust cuticle involved in the formation of cross-links from *N*-acetyldopamine. J. Insect Physiol. 18, 527—540.
2. Andersen, S.O. (1973) Comparison between the sclerotization of adult and larval cuticle in *Schistocerca gregaria*. J. Insect Physiol. 19, 1603—1614.
3. Andersen, S.O. (1974) Cuticular sclerotization in larval and adult locusts, *Schistocerca gregaria*. J. Insect Physiol., 20, 1537—1552.
4. Andersen, S.O. and Barrett, F.M. (1971) The isolation of ketocatechols from insect cuticle and their possible rôle in sclerotization. J. Insect Physiol. 17, 69—83.
5. Hepburn, H.R. and Joffe, I. (1974) Hardening of locust sclerites. J. Insect Physiol. 20, 631—635.
6. Andersen, S.O. (1974) Evidence for two mechanisms of sclerotization in insect cuticle. Nature, 251, 507—508.
7. Pryor, M.G.M. (1940) On the hardening of the cuticle of insects. Proc. R. Soc. London Ser. B 128, 393—407.

Comparative Physiology — Functional Aspects of Structural Materials
Eds L. Bolis, H.P. Maddrell and K. Schmidt-Nielsen
© North-Holland Publishing Company — 1975 — Amsterdam

Mechanical instabilities in biological membranes

J.E. GORDON

University of Reading, Department of Applied Physical Sciences, Reading RG6 2AY (Great Britain)

INTRODUCTION

The engineer has always been accustomed to choose his structural materials for traditional and pragmatic reasons. Very few engineers have given much thought to the fundamental philosophy which links materials to structures, partly because most of them do not see that such a philosophy is practicable or necessary and partly because, except at a trivial level, the subject is intellectually very difficult. At present almost no useful body of tested and established thought can be said to exist in this field.

This may or may not matter in engineering but it is likely to be important in biology where the relation between elasticity and structure seems to have been ignored as irrelevant or incomprehensible until recently. However, there is no reason to suppose that while nature is infinitely subtle in her chemistry and cybernetics she should not be equally subtle where structures are concerned. Even the simplest plants and animals are subject to various mechanical loads and unless they are structurally adequate they can neither exist nor come into existence. In the simpler forms of life nature has not made much use of pre-existing inorganic solids for structural purposes but has apparently preferred to evolve, de novo, organic loadbearing materials especially suited to the purpose.

The purpose of this paper is to ask some very, very simple but, I think, fundamental questions about the mechanical functions and elastic requirements of living membranes. The method employed is purely a "black-box" approach and no assumptions are made about molecular mechanisms.

THE PROBLEM OF CONTAINING A FLUID

The engineer usually arranges to confine gases, liquids and slurries within

50

more or less rigid containers such as tanks, pipes and boilers. The material of these containers is generally assumed to operate only at small strains and to obey Hooke's law. Within these limitations, the behaviour of containers under internal pressures has been extensively studied and is generally held to be fairly well understood although such vessels do, in fact, burst quite frequently.

Hookean materials of the engineering kind are generally limited in practice to strains of less than 1% and thus seem to be ruled out for use in biological membranes. It is important to realise, however, that most of our preconceptions about pressure vessels are based upon the implicit assumption that the material of the walls more or less obeys Hooke's law (Fig. 1a). Once this condition is abrogated the nature of the problem changes dramatically.

Thus in the engineering elasticity which operates at small strains we are archetypally concerned with Hookean behaviour in which stress is proportional to strain or nearly so. In biological soft tissues we might almost say that we are archetypally concerned with an elasticity in which stress is independent of strain (or nearly so) up to a certain degree of extension

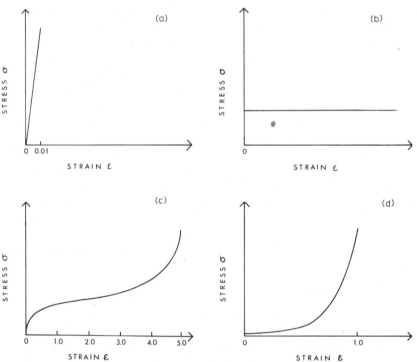

Fig. 1. Stress-strain curves for various kinds of materials. (a) Hookean material where $F = A\sigma = AE\epsilon$ and where $\epsilon_{max} < 0.01$. (b) A material in which force is independent of strain. $F = A\sigma$ and ϵ_{max} is indefinite. (c) Rubbery material where $F = A\sigma = A(\alpha\epsilon^n + \beta\epsilon^m)$ where α and β are positive, $n \approx 2$, $m \approx 1/4$ and $\epsilon_{max} \approx 5.0$. (d) A typical multicellular membrane. $F = A\sigma + A(\gamma + \theta\epsilon^n)$ where $n = 2-3$ and $\epsilon_{max} \approx 1$.

(Fig. 1b). In such an elasticity the resistance to extension (and thus the stability of the structure) is not, in general, due to an increase of force with strain but to an increase of strain energy. (Strain energy is the integral of stress times strain.)

Such behaviour is, of course, exhibited by the surface tension of a liquid and it is possible that simple surface tension may have been the predominant form of elasticity in the most primitive kinds of life and may well play a role in present-day biological membranes though I am of course aware that the elasticity of most modern living membranes is complex.

Membranes exist for the purpose of carrying loads in two dimensions. To study them uniaxially may be irrelevant or deceptive. The wall of a pressure vessel is subject to two principal stresses in its own plane. For our immediate purpose we may perhaps neglect the third principal stress normal to the surface.

So long as we are operating in a region of the stress-strain diagram in which stress is independent of strain then the membrane cannot resist shearing forces in its own plane. This is clearly true of surface tension but it must also be true of all other mechanisms which produce the same effect. Furthermore, Poisson's ratio has no meaning.

The simplest biological container is presumably a globular or spherical one and here the two principal stresses operating in the plane of the membrane are equal and so perhaps no serious problem arises. Once we need to construct an elongated animal we are, however, faced with the problem of maintaining the equality of the principal stresses if we want to continue to use a material where stress is substantially independent of strain. This can be done in two ways:

(1) By making the cuticle of the animal from spherical surfaces. Thus we end up with a segmented shape. In view of the substantial advantages in other connections of a material whose stress is strain independent this fact may have a bearing on the frequency with which segmented forms occur in worm-like animals.

(2) Since in an ordinary cylindrical pressure vessel the circumferential stress must be twice the axial stress equality of principal stresses may be restored by the application of an additional longitudinal force which must, however, be reacted externally and this is not practicable in a very simple animal. Such a system is, however, characteristic of parts of advanced animals (e.g. arteries in the vertebrates) where the additional tension can be reacted, in the end, by compressions in the bones.

If neither of these solutions is practicable then nature is compelled to abandon a membrane tissue in which stress is truly independent of strain. For such a material the stress-strain curve must sooner or later trend upwards (Fig. 1d). That is to say stress must increase with strain. However, the advantages of a curve with what might be described as a "flat bottom" are great and we must consider the problems of an elasticity of this kind.

TYPES OF ELASTICITY

The load in a simple surface-tension system is characterised by:

$$F = l\Gamma$$

where F = force or load, l = width with which we are concerned, Γ = surface tension per unit width.

Here F is independent of both the extension and also of the thickness or cross-sectional area of the material. The size of an animal using this type of elasticity is strictly limited since F cannot be increased by making the material thicker.

In most biological soft tissues at small strains it is more nearly true to write:

$$F = \sigma l t = \sigma A$$

where σ = stress and t the thickness of the material (so that $lt = A$, the cross-sectional area). A molecular mechanism by which this effect can be realised has been suggested recently by J. Gosline (personal communication) and this or some equivalent mechanism represents a very important step in biomechanical evolution.

As we have said the development from a globular to a tubular animal requires an upward trending stress-strain curve. With this will come the ability to resist shear, the acquisition of a Poisson's ratio and so forth. The most familiar strain-dependent elasticity is the Hookean one where:

$$F = A\sigma = AE\epsilon$$

where ϵ is the strain. E is the Young's modulus (i.e. $E = \delta\sigma/\delta\epsilon$ = Const.)

When ϵ is small this relation leads of course to the classical Hookean elasticity of the engineer. Such an elasticity does not, however, hold true at large strains. In any case not many materials obey Hooke's law at strains of more than a few percent. Tendon does approximate to Hooke's law up to strains of 10—15% but tendon is loaded uniaxially and is not used for membranes.

Elastomers of a rubbery nature usually possess stress-strain curves which may be generalised in the form:

$$\sigma = \alpha\epsilon^n + \beta\epsilon^m$$

This equation is capable of representing a wide range of elasticities where the stress-strain curve passes through the origin. For rubber-like materials:

$\alpha = +ve$

$n > 1$ (typically about 2)

$\beta = +ve$

$m < 1$ (typically $\frac{1}{4} - \frac{1}{2}$)

Thus we get a typical rubbery "sigmoid" curve. (Fig. 1c).

Suppose a tubular pressure vessel, such as an artery, to be made from such a material. The classical engineering formula for the stress in a thin-walled cylinder under pressure is:

$$\sigma_c = \frac{Rp}{t}$$

where σ_c = circumferential stress, R = radius of vessel, t = thickness of wall, p = internal pressure.

But this no longer holds at large strains and must now be written*:

$$\sigma_c = \frac{R(1 + \epsilon) p}{t} = \alpha\epsilon^n + \beta\epsilon^m \qquad \text{Thus } p = \frac{t}{R} \cdot \frac{\alpha\epsilon^n + \beta\epsilon^m}{(1 + \epsilon)}$$

The relation between pressure and strain is thus:

$$\frac{\delta p}{\delta\epsilon} = \frac{t}{r} \cdot \frac{\alpha\epsilon^n \left(\dfrac{n}{\epsilon_1} + n - 1\right) + \beta\epsilon^m \left(\dfrac{m}{\epsilon} + m - 1\right)}{(1 + \epsilon)^2}$$

The tube is stable under pressure so long as $\delta p/\delta\epsilon$ is positive but if and when $\delta p/\delta\epsilon \leqslant 0$ the system becomes unstable and will bulge into an "aneurism". Thus when $m = 1/2$ system is unstable for $\epsilon > 1$ and when $m = 1/4$ system is unstable for $\epsilon > 1/3$. But these values of m and ϵ are typical of rubbery elasticities and of biological strains. The instability of a thin rubber tube (i.e. its tendency to produce "aneurisms" under internal pressure) is easily demonstrated. It follows that an elasticity of this type is unsuited for many biological purposes and does in fact appear to be fairly rare in nature. Thus

---

*Here, and in what follows, I am using the "pragmatic" values of stress. That is to say, stress values based upon the original cross-section and not corrected for reduction in area with increase of strain. If this is borne in mind and used consistently no error is introduced. Nearly all of the published biological stress-strain curves are of this type. Usually the adoption of this convention somewhat simplifies the resultant mathematics.

biomechanical evolution by means of ordinary flexible polymer chains of the rubbery kind is not very promising.

The elasticity of real soft tissues is more likely to be represented by an equation of the type:

$$\sigma = \gamma + \phi\epsilon^n$$

Where γ is the "surface tension" or strain independent term, ϕ is a constant corresponding to a "modulus" and $n > 1$, and is perhaps usually from 2 to 3. Thus in a tubular vessel:

$$p = \frac{t}{R}, \frac{(\gamma + \phi\epsilon^n)}{(1 + \epsilon)}$$

Whence, for instability:—

$$\epsilon^{(n-1)}(n + \epsilon(n-1)) < \gamma/\phi$$

The linear case where $n = 1$ hardly seems to occur at large strains but it may be worth noting that the system is stable when $\gamma/\phi < 1$ for all values of ϵ. Thus Hooke's law ($\gamma = 0$) is stable at all strains. For the more realistic values, for instance where $n = 2$ then the system is stable where:

$$\epsilon < -1 \pm \sqrt{1 + \gamma/\phi}$$

So to be stable up to $\epsilon = 1.0$, $\gamma/\phi < 3$.
Similarly whn $n = 3$ and $\epsilon = 1.0$ system is stable when $\gamma/\phi < 5$.

Thus there is a requirement to keep v small and n relatively large and this in fact corresponds pretty closely to the general shape of the stress-strain curves of many of these materials.

WINGS AND FINS

Where we have to deal with low structure loading coefficients*, which is the case with the wings and fins of animals and also with sailing ships, kites, tents and slow aircraft, it is easy to show that the most efficient way to provide a strong impermeable surface (e.g. a wing) is to use a thin membrane stretched over a "rigid" framework. Resistance to normal fluid pressure is provided by the curvature of the membrane whose tension is converted into compression in the supporting spars. Fracture, when it takes place, usually occurs by the breaking of the "rigid" sub-structure. With engineering artifacts not only the supporting spars but also the membrane covering are generally of a Hookean character. It follows that, if the modulus of the membrane is relatively high, as is usually the case, then the load in the supporting framework is nearly proportional to the aerodynamic or hydrodynamic pressure which again is proportional to the square of the fluid velocity. Thus destruction is easy.

Here we encounter one of the many advantages of the type of membrane elasticity which we have been discussing. Many biological surfaces such as the wings of a bat or a pterodactyl or the fins of a fish consist of membranes which are supported by bony frameworks which appear to the eye of the engineer to be quite inadequate in thickness and in stiffness and, moreover, to be made from relatively brittle material. However, the membrane is enabled to bow or belly out between the supports so that an increase in aerodynamic pressure is accommodated with little increase in the load upon the supporting bones, the increased pressure being met by a decrease in radius of curvature. Photographs of a fruit-bat in flight, for instance, show that on the down-beat of the wing the profile of the wing membrane approaches a semi-circle.

The extension of a membrane from a flat to a semi-circular shape requires a strain of 0.57. Such an extension lies typically almost within the "flat" region of the stress-strain curve of many membrane materials and, as we have seen, provided certain conditions are fulfilled, the extension is a stable one.

STRESS CONCENTRATIONS

Besides the gross instabilities to which they are potentially subject biological materials are liable (like all other materials) to simple fracture although in

---

* The weight efficiency of a structure can be analysed parametrically. The subject is complex but broadly speaking, the weight of a structure will tend to be a function of the magnitude of the load to be sustained and the length over which this load has to be carried. Thus, for instance, if a load P is carried in compression in n panels over a distance L then efficiency

$$\eta = \frac{W}{P} = \text{Const.} \times n^{\frac{2}{3}} \left(\frac{\rho}{\sqrt[3]{E}} \right) \frac{(L^{\frac{5}{3}})}{(P^{\frac{2}{3}})}$$

where η = weight efficiency of structure, W = weight of structure, ρ = density of material, E = Young's modulus, $L^{5/3}/P^{2/3}$ = a structure loading coefficient $\rho/\sqrt[3]{E}$ = a material efficiency criterion.

Comparable, but algebraically different, relationships exist for the various structural cases. Broadly speaking, where compression and bending are concerned, there is a very severe relative weight penalty in carrying a light load over a long distance. The tension cases are, however, relatively insensitive to P/L. It follows that for lightly loaded structures such as most animals (especially flying animals) it will pay to use kite-like structures having a membrane supported upon the minimum of bending and compression framework.

real life this does not very often seem to happen. In fact, of course, the mechanics of fracture in solids is never simple and the subject is a difficult one.

Hookean solids seldom break as a direct consequence of the mean stress which is applied to them. Unless the geometry of the solid is quite remarkably uniform every local irregularity of shape (or of elasticity) will cause a local increase or concentration of stress. For a simple crack in a Hookean solid the stress is raised locally by a factor K where:

$$K = 1 + 2\sqrt{l/r}$$

where l = half length of crack,
r = radius of crack tip.

K may easily reach a value of 100 or more and it is this local stress which usually precipitates mechanical fracture.

For a material where stress was wholly independent of strain it is clear that concentrations of stress could not occur. In real biological tissues although stress is not strictly independent of strain the shear modulus over the lower or working part of the stress-strain curve must be so low that the great majority of the stresses in the material cannot be effectively concentrated at geometrical discontinuities. In this characteristic such materials are directly opposed to the behaviour of rubber, for instance, which is highly sensitive to stress concentrations as is easily shown by pricking a balloon.

If a hole or a crack should occur, the low modulus and high strain prevailing in most membranes must result in rounding off the crack tip so as to produce a large radius.

GRIFFITH CONDITIONS

Fracture is not governed solely by force considerations. Griffith stated that a crack can only propagate when the release of strain energy caused by the relaxation of the stressed material behind the fracture surfaces is sufficient to "pay" for the work of fracture of these newly created surfaces. Thus for a Hookean solid the critical Griffith crack length l is given by:

$$l = \frac{2W}{\pi E \epsilon^2}$$

where W = work of fracture, E = Young's modulus. For engineering solids W may be in the region of 10^7 ergs/cm^2 and E about 10^{12} dynes/cm^2.

Thus:

$$l \approx \frac{2 \cdot 10^7}{\pi \cdot 10^{12} \cdot \epsilon^2}$$

So, if we want $l \approx 10 \, \text{cm}$ then $\epsilon \approx 10^{-3}$, which is typical of engineering strains. For most ordinary "hard' solids

$$\frac{\Gamma}{E} \approx 10^{-9} \, \text{cm}.$$

So if $W \approx \Gamma$ and $\epsilon \approx 0.5$, then $l \approx 4 \cdot 10^{-9}$ cm, which of course is absurd. To achieve even $l \approx 1.0$ cm we should need $W \approx 10^8 - 10^9$ ergs/cm^2 which is as much as steels and impracticable in biological materials. It is therefore necessary to reduce E drastically.

If $\Gamma \approx 100$ ergs/cm^2 and the material were frozen then E would be roughly 10^{11} dynes/cm^2 (10^5 Kg/cm^2 = 10^4 N/mm^2 = $1.4 \cdot 10^6$ lb/inch^2.)

In fact, the tangent modulus of unfrozen typical soft tissues at 0.5 strain is often between 1 and 2 Kg/cm^2 or about 10^6 dynes/cm^2. Thus a critical Griffith crack length of 1 cm now requires a work of fracture in the region of $2 \cdot 10^5$ ergs/cm^2 which is a possible and indeed a probable figure.

CONCLUSIONS

The majority of biological soft tissues which operate at high strains have very typical (and generally similar) stress-strain curves which differ greatly from those of practically all inorganic materials and man-made polymers. At first sight the shape of these curves seems irrational and possibly accidental. In fact this is very far from being the case. It has been shown, I think, on several grounds that this shape is virtually the only shape which is safe or practicable for a membrane material operating at high mechanical strains. Far from being fortuitous the evolution of tissues with this particular shape of stress-strain curve seems to be an essential condition for the development and continuing existence of living creatures as we know them. The development of a molecular mechanism capable of giving just this type of relationship between stress and strain was therefore a matter of great importance.

PART 2

CONTRACTILE MECHANISMS

Comparative Physiology — Functional Aspects of Structural Materials
Eds L. Bolis, H.P. Maddrell and K. Schmidt-Nielsen
© North-Holland Publishing Company — 1975 — Amsterdam

Bacterial flagella: structure and function

L.M. ROUTLEDGE

Agricultural Research Council, Department of Zoology, University of Cambridge, Cambridge CB2 3EJ (U.K.)

INTRODUCTION

The flagella of bacteria are fundamentally different from those of higher organisms. Though both types of flagella can have a helical waveform, the ultrastructure and mechanisms of movement are in no way comparable. The complex flagella of eucaryotes have a characteristic 9 + 2 arrangement of microtubules and bending movements are produced by sliding of these tubules over each other [1,2].

Bacterial flagella on the other hand do not have the 9 + 2 arrangement. They consist of a single protein polymerised to form a tubule or flagellar filament. In some bacteria (e.g. *Vibrio metchnikovii*) this filament is enclosed in a polysaccharide sheath. The arrangement of flagella on bacteria varies, some (e.g. *the rod shaped Pseudomonads*) have a single polar flagellar filament, others (e.g. *Spirillum*) have polar bundles of flagella. Another group (*Salmonella, Bacillus*), the peritrichous flagellated bacteria, have filaments over the whole cell surface which can also form bundles. It is these bundles of flagella filaments which can be seen as a helical tail in dark-field microscopy and occasionally in fixed bacteria (Fig. 1). Under dark field microscopy a wave appears to travel down this bundle in the bacterium which can swim at speeds of up to 40 μm/s. This wave appearance could be produced by (1) a helical bending wave or (2) a rotation of rigid helical filaments. These two hypotheses for flagella movement are dominant at present.

The implications of both of these hypotheses are startling. If a bending wave passes down the flagella then molecular changes must be propagated down the structure. If, however, flagella rotate then they must have a rotary motor: and one could even suggest that Nature must indeed have invented the wheel.

With the introduction of the electron microscope great advances have been possible in the area of flagella structure which give indications of the mechanism by which bacteria swim.

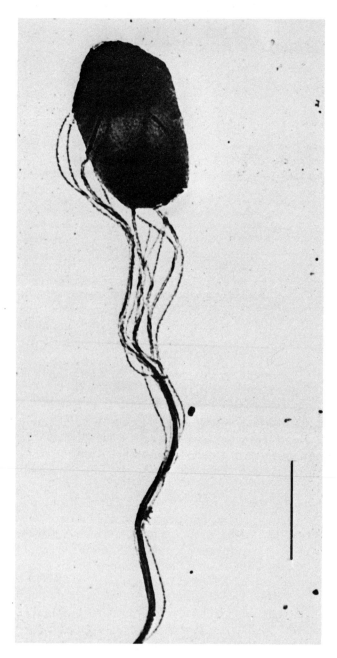

Fig. 1. Electronmicrograph of *Salmonella abortus-equi* with flagella forming a bundle or tail of characteristic waveform. Specimens were negative-stained with 1% uranyl acetate (pH 4.3). Bar = 1 μm.

FLAGELLAR ULTRASTRUCTURE

Bacterial flagella have three distinct regions which can be recognized in the electron microscope. These are: a basal complex within the bacterium, a hook region which penetrates the cell wall, and a helical filament of simple construction [3]. A diagrammatic representation of these structures is shown in Fig. 2 which has been compiled from electron micrographs of negatively stained and sectioned material [4—6]. The majority of studies on bacterial flagella have investigated the filament as this is more than 90% of the organelle and is clearly visible. More recently the basal disc complex and the hook have been studied and these are of great interest as possible power-generating structures.

BASAL COMPLEX

Flagella with intact basal structures can be prepared from bacteria by first removing the cell wall with lysozyme and the chelator EDTA. The resulting spheroplasts are lysed in the detergent Triton X-100. The flagella with attached hook and basal complex prepared by this procedure are free from cytoplasmic membrane which may, however, be present in those prepared from partially degraded cells of *Proteus vulgaris* [3] and lysed *Rhodospiril-*

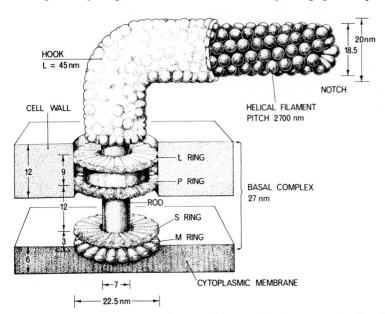

Fig. 2. A generalised reconstruction of the proximal portion of a flagellum from a gram-negative bacterium, showing the basal complex, the hook, and part of a helical flagella filament. (Modified from DePamphilis and Adler [6].)

lum [4]. A generalized model for the basal end of a flagellum of a gram-negative bacterium is shown in Fig. 2. The basal structure consists of a rod on which are mounted two pairs of rings, the L- and P-rings associated with the bacterial cell wall and the S- and M-rings near the cytoplasmic membrane. The L- and P-rings are associated with the lipopolysaccharide and peptidoglycan layers in the multilayered cell wall of gram-negative bacteria. DePamphilis and Adler [5] also suggested that a collar exists between the L- and P-rings as negative stains do not penetrate between these rings which are separated by 9 nm. This pair of rings appears to be absent in the gram-positive bacteria which also lack a multilayered cell wall [5]. Therefore the L- and P-rings may not be a part of the motility mechanism.

A lower pair of rings, the S- and M-rings, are present in both the gram-negative and gram-positive bacteria. These rings are separated by only 3 nm yet stain penetrates between them, and the M-ring, which is associated with the cytoplasmic membrane, can be easily detached from the basal complex [6].

The M-ring consists of sixteen segments and thin sections show it to be partly embedded in the cytoplasmic membrane [6]. The S- or supramembrane ring differs from the other rings in that it does not seem to be associated with a membrane and the S-ring is not in register with any known cell envelope structure. DePamphilis and Adler [6] also investigated ATPase activity in flagella and membrane fractions. They found no ATPase activity in flagellar filaments, hooks, basal bodies. They did however detect strong ATPase activity in the cytoplasmic membrane though there was no correlation between this activity and the sites of attachment of flagella. These findings agree with those of earlier studies which indicated a complete absence of ATPase activity from flagellar filaments [8,9].

The basal complex is assumed to be the force generating structure in both current hypotheses of flagellar movement and it is important to know how energy is utilized by this structure. It has not been possible to study the metabolism of isolated basal bodies but Adler and his collegues [10] have shown that ATP is required for chemotaxis but not motility in *Escherichia coli* and an as yet unidentified intermediate of oxidative phosphorylation is required for flagella movement.

THE HOOK

A curved segment connects the basal complex to the flagellar filament, this is termed the hook region. The hook is a tubule between 40 and 90 nm long and has a slightly larger diameter than the filament region [11]. Koffler and his co-workers [12] have isolated hooks. They are composed of a protein of molecular weight 33 000 and are antigenically distinct from the filament protein [13]. The hook is more resistant to heat dissociation than the filament. The arrangement of the protein subunits in the hook is not understood, since the great flexibility of the hook creates difficulties in the use of optical filtering techniques which have been used extensively to examine the filament structure. The protein subunits are smaller than those of the filament, yet the hook has a greater diameter than the filament. This suggests that the number of subunits in a cross-section of the hook is greater than of the filament. The subunits of the hook are elongated and lying at an angle in the tubule as can be deduced from the "notched" appearance of hooks in negatively stained electron micrographs (i.e. [6]: Fig. 24). The flexibility of the hooks allows the flagella to form a bundle as shown in Fig. 1. Andersen [14] has shown that with increasing fluid viscosity the flagellar bundle lies more nearly in the major axis of the cell suggesting that there is a balance between the hydrodynamic forces tending to align the bundle and elastic forces associated with the bending of the flagella around the sides of the cell.

Bacterial mutants which are unable to control the hook length have been isolated [15]. The "polyhooks" are 1—2 μm long and are arranged in tight helical coils of pitch 0.14 μm. These cells lack translational motility, though experiments in which the polyhooks are tethered have shown that the cells rotate [16]. When polyhooks are bound together by antihook antibody they may be straightened (suggesting that they are normally somewhat flexible) [16].

THE FILAMENT

The flagellar filament has been well characterized and its structure is known, at least in *Salmonella*. The pioneering work of Weibull [17] on *Bacillus* and *Proteus* showed that the flagellar filaments were polymers composed of protein monomers of molecular weight 40 000. This protein was given the name flagellin and has an unusual amino acid composition [18]. Flagellar filaments can be dissociated into flagellin by heat, acid, urea and other denaturants, and can be induced to reassemble to form filaments morphologically indistinguishable from the original organelle [19,20].

The shape of a flagellar filament is helical, but the helix becomes flattened to a near-sinusoidal wave in negatively stained preparations made for electron microscopy (see Fig. 1). The wavelength varies from species to species but is generally between 1—3 μm [21]. In *Salmonella* the wavelength of the wild type is 2.5 μm mutants with half-normal wavelength (1.2 μm) and with straight flagella have been isolated [22,23]. The wild-type cells undergo rapid translational motion of up to 40 μm/s. The half-normal or Curly mutant shows rotational tumbling movements and the mutant with straight flagella is non-motile. The transition from normal to curly waveform has been observed in flagella filament bundles and has given rise to speculation that this was part of the propulsion mechanism [24]. The transition from normal to curly waveform can be induced by changes in pH [25] and by a single amino acid substitution [22]. The transition from normal to straight flagella is due to a single amino acid substitution in *Salmonella* [23] and in *Bacillus subtilis* [26] this is known to be the substitution of valine for alanine. Thus the waveform of the flagella has an inherent and specific variability. This fact was emphasised by Asakura and Iino [27]. When flagellins from normal and straight flagella are polymerized together at different ratios, three intermediate waveforms are produced. Thus flagellins have the ability to pack in several distinct ways to form the flagellar filament.

The arrangement of flagellin in the filament has been examined by X-ray diffraction [28,29]. Electron microscopy [30,31] and by optical filtering of electron micrographs [32–34]. The earlier studies showed that the subunits repeat every 5.2 nm and that they have near-hexagonal surface packing in the tubular filament. Due to difficulties of specimen orientation for X-ray diffraction and the superposition of back and front images in electron microscopy, many different models have been presented for the filament ultrastructure [31,32,35,36].

The use of optical filtering and tilting of electron micrographs [33,34,37] has allowed separation of the front and back images of the flagellar filament. The number of strands in cross-section is odd, with the most likely number as eleven. The filament in *Salmonella* and *Pseudomonas* is packed as shown in Fig. 2 and can be imagined as being a five, a six or an eleven start helix and can be classed as a (5,6,11) structure in the nomenclature of Erickson [38]. Cross-sections of flagella do not show the subunits clearly though with rotational enhancement 10 or 11 subunits are visible [39].

When fragments of flagella filament are observed in the electron microscope they have structural polarity, one end having a "notched" or "fish tail" appearance and the other end being rounded [35,40,41]. This "notch" appearance is due to the elongated flagellin molecules lying tilted in the filament (see Fig. 2) producing a funnel which becomes filled with negative stain in electron microscopy [35,39,42].

The flagellar filament and hook have a hollow core of between 2.9 and 6.5 nm diameter [30,35,39,41,43]. Sleytr and Glauert [44] using freeze etched flagella from *Clostridium* showed that the core contained no structural material and this agrees with the penetration of negative stain into short fragments of flagella [43].

The "notch" is at the tip or the distal end of the flagellar filament [39,40] and it is from this end that flagellar growth takes place both in vitro [45] and in vivo [46,47]. Flagellin is synthesised on ribosomes within the cell [48] and several mechanisms for its transport to the distal end of flagella have been considered [49]. Flagellin may be secreted into the medium, travel down the outside of the filament, or tunnel down the central core. Sensitive antibody fixation tests have been unable to detect free flagellin in the medium but little direct evidence is available to examine the other mechanisms of filament growth. Bacteriophages which attack flagella are thought to move down the helical grooves of filaments [50] similarly the flagellin molecules could travel along these helical grooves from the proximal to the distal end of the filament but this seems unlikely. A more likely explanation for filament growth, is that flagellin tunnels down the central core and this may be facilitated by the electrical gradient along the filament discribed by Gerber and his co-workers [42,51]. The electrical gradient may act on the negatively charged flagellin molecules to produce a propelling force.

Flagellin filaments can be reconstituted onto bacteria [52]. Non-motile *Salmonella* with stumps of straight flagella were incubated on solutions of normal waveform flagellin. Filaments of normal waveform polymerised on the distal end of the straight flagella and the cells became motile. In order to get significant incorporation of externally added flagellin, the endogenous synthesis of flagellin had to be inhibited; flagellin from the cell competed significantly with externally added flagellin even when this was added as a highly concentrated solution. This suggests that the endogenous flagellin was also at a high concentration (> 10 g/l), perhaps due to its compartmentation in the flagellar core.

THE MECHANISM OF MOTILITY

Propagated Helical Wave

This mechanism was first suggested by Bütschli [53] and developed by Reichert [54]. They considered a rod with a "line of contraction" helically wound round its surface. If this helical line shortens relative to other parallel helical lines, then a large scale helix is produced. If this line of contraction is transferred to the next parallel helical line then the effect is to propagate the wave along the rod. This hypothesis was somewhat supported by the discovery that the flagellin molecules in the flagella are arranged in helical rows [30] (see also Fig. 2). From radial projections of straight and helical flagella it can be calculated that the shortening of this line of contraction would be between two and four percent or between 1 and 2 Å per flagellin molecule [36,41]. A comparison of straight and normal flagella [33] shows a change in the tilt of the long pitch helical rows from 7° in the straight filament to less than 2° in the normal waveform. This they suggest is due to changes in the subunit orientation in straight and normal flagella. A change in the orientation of the elongated flagellin molecule would produce a change in the dimension subtended at the surface of the filament. The effect of this would be to increase or decrease the length of the helical row to which the molecule belongs. An important question in this hypothesis is how is this change produced and propagated? An analogy may be drawn to the 'contraction' of T-even phage sheath [55]. Here again the protein subunits are helically arranged, the contraction is initiated and a wave of contraction passes along the sheath. This contraction is due to the rearrangement of the subunits in a displacive manner, with the formation of a different pattern of inter-subunit bonding. The 'extended' sheath is the unstable form and energy is released on going to the 'contracted' state. A small change in the conformation of the subunits is also likely, though the major movement is due to sliding changes in subunit contacts. Reversible conformational changes have been noted in flagellin [42,55,55a] with a small change in the axial ratio of the molecule. In flagellar filaments [42,51] a change in conformation occurs at $30°C$ which may indicate a structural rearrangement. In the T-even phage sheath the wave of contraction is irreversible: the contracted form is permanently at a lower energy level. A propagated wave in a flagellar filament must be reversible; the flagellin molecules reverting to their 'extended' state after the bending wave has passed. This has given rise to the idea of two quasi-equivalent states of conformations for flagellin molecules in the flagellum [41,57]. In this theory, the energy for a helical bending wave derives from movement of the hook-basal complex and induced conformational or orientation changes in the flagellin molecules. The bending wave may be propagated in an allosteric fashion down the flagellum from subunits in one helical row to the next helical row.

Rotation of Rigid Helical Filamenta

The rotation of rigid helical filaments relative to the bacterial cell body was suggested as a possibility by Stocker [58] and Doetsch [59]. Mussill and Jarosch [60] observed that in *Spirillum* the polar flagellar bundle can be held stationary and the cell body appeared to rotate about the point of flagellar bundle insertion. Berg and Anderson [50] have suggested, that "each filament rotates individually". The clockwise (relative to the cell) rotation of a rigid left-handed helix will produce forward propulsion, much like an Archimedean screw. This has been analysed hydrodynamically [61]. The hydrodynamic analysis cannot distinguish propulsion by a propagated wave or by rotation of a bundle of filaments [62]. If individual filaments are rotating in a flagellar bundle how do they not become entangled? If left-handed helical filaments rotate clockwise a point of entanglement in adjacent flagella will travel towards the free ends of the flagella [36]. For anti-clockwise rotation however, a point of entanglement will travel to the bases of the filaments and prevent further rotation.

The evidence. in favour of the rigid rotor model for bacterial flagella motility are: the structure of the basal complex, the effect of cross linking flagella filaments, and the rotation of tethered bacteria.

The basal complex (Fig. 2) though suggestive of a rotary motor [50] could perform almost any function required in a theory of flagella motility.

Several motility mutants have been examined for structural changes in the basal complex without success [5].

Peritrichous flagellated bacteria and bacteria with a single flagellar filament are motile at high concentrations of univalent antibody [63]. However, at low concentrations of divalent antibody, peritrichous bacteria are immobilized, whereas bacteria with a single flagellum are not. This observation has been used as evidence for rotating filaments [50] on the basis that, fixed points on the filaments move relative to each other in a rotation model, but not in a propagated wave model. Thus divalent antibodies would inhibit adjacent rotating filaments but not filaments propagating a bending wave. From a consideration of lateral movements of adjacent filaments, Calladine [64] showed that even in a propagated wave, large relative displacements of filaments occur which would be blocked by divalent antibodies. Similarly the binding of a single flagella-specific phage prevents motility in peritrichous bacteria and has been used as evidence for rotating filaments; the large phage blocking the relative rotation of filaments [7]. Considering bundle formation however, it can be seen that binding of adjacent filaments, by a phage or divalent antibody, will prevent entanglements of flagella from passing to the tip of the filaments. Therefore both mechanisms of motility predict that the peritrichous bacteria will be immobilised by the crosslinking of the filaments.

The strongest evidence for rigid flagella rotation comes from tethered bacteria. Bacterial mutants with polyhooks and without flagella filaments are normally non-motile. If, however, they are attached to microscope slides using antihook antibodies they can be observed to move [10,16]. It is not easy to discern whether the cells rotate or merely move in an arc around the tethered filament. The interpretation of these authors is that the cells normally move in an anticlockwise direction at 2—9 rev./s for several revolutions but also appeared to reverse and turn clockwise also for ten or more revolutions. The tendency to turn clockwise increases when repellents are added and it has been suggested that this is the basis of the chemotactic response [65]. Normally bacteria move in straight lines then 'twiddle' and change direction [66]. In a favourable gradient the runs before "twiddling" are longer. The net result is to produce a bias for movement in the favourable direction, the attractant leading to an inhibition of "twiddling". The "twiddling" behaviour is due to the pulling apart of the flagellar bundle and its reformation over a different region of the cell surface. Reversal of flagella movement would tend to destabilise the flagella bundle and could be responsible for the observed "twiddle" behaviour.

Where two polyhook mutants are bound together with antihook antibodies they appear to turn relative to each other [16]. This is strong evidence that bacteria may rotate their flagella. In the propagated wave hypothesis only limited rotation of the filaments relative to the cell body is predicted. The contraction of T-even phage sheath leads to a rotation of the capsid of almost 360° relative to the base plate [55]. However, the transfer of a line of contraction around the flagella filament would not result in a net rotation of the filament.

When normal waveform flagellin is polymerised onto straight flagella of the non-motile mutant, the bacteria become motile [52]. The propagated wave hypothesis would not predict motility in these cells unless a signal can be transmitted to the normal flagella through the straight flagella segment. According to Silverman and Simon [16] tethered mutants with straight flagella can rotate. If this observation is correct then the propagated wave hypothesis as presently understood, is untenable. Further observations are necessary to establish these important points.

It seems likely that future research will be directed particularly to the structure and biochemical properties of the basal complex. It may also become possible to make direct observations of the flagellar waveform by electron microscopy of hydrated or even living bacteria.

CONCLUSION

The structure of the flagellar filament, hook and basal complex is now understood to a reasonable degree. The opposing hypotheses for flagellar motility, namely propagation of a helical bending wave or rotation of a rigid helic cannot yet be resolved with absolute certainty. Further experimental studies are required in order to distinguish between the hypotheses. These studies will repay careful examination however as bacterial flagella have unusual features which have implications in many areas of biology.

REFERENCES

1. Satir, P. (1965) Studies on cilia II. Examination of the distal region of the cillary shaft and the role of the filaments in motility. J. Cell. Biol. 26, 805—834.
2. Summers, K.E. and Gibbons, I.R. (1971) Adenosine triphosphate-induced sliding of tubules in trypsin-treated flagella of sea-urchin sperm. Proc. Natl. Acad. Sci. U.S. 68, 3092—3096.
3. Abram, D., Koffler, H. and Vatter, A.F. (1965) Basal structure and attachment of flagella in cells of *Proteus vulgaris*. J. Bacteriol. 90, 1337—1354.
4. Cohen-Bazire, G. and London, J. (1967) Basal organelles of bacterial flagella. J. Bacteriol. 94, 458—465.
5. DePamphilis, M.L. and Adler, J. (1971) Purification of intact flagella from *Escherichia coli* and *Bacillus subtilis*. J. Bacteriol. 105, 376—383.
6. DePamphilis, M.L. and Adler, J. (1971b) Fine structure and isolation of the hook-basal body complex of flagella from *Escherichia coli* and *Bacillus subtilis*. J. Bacteriol. 105, 384—395.
7. Berg, H.C. (1974) Bacterial Movement, in Bacterial Locomotion. Abstr. Symp. Swimming Flying Nat. Calif. Inst. Tech., p. 3.

8. Barlow, G.H. and Blum, J.J. (1952) On the 'contractility' of bacterial flagella. Science 116, 572.
9. Newton, B.A. and Kerridge, D. (1965) Flagellar and ciliary movements in microorganisms. Symp. Soc. Gen. Microbiol. 15, 220—249.
10. Larsen, S.H., Reader, R.W., Kort, E.N., Tso, W. and Adler, J. (1974) Change in direction of flagellar rotation is the basis of the chemotactic response in *E. coli*. Nature 249, 77.
11. Lowy, J. (1965) Structure of the proximal ends of bacterial flagella. J. Mol. Biol. 14, 297—299.
12. Abram, D., Mitchen, J.R., Koffler, H. and Vatter, A.E. (1970) Differentiation within the bacterial flagella and isolation of the proximal hook. J. Bacteriol. 101, 250—261.
13. Dimmit, K. and Simon, M.I. (1971) Purification and partial characterization of *Bacillus subtilis* flagellar hooks. J. Bacteriol. 108, 282—286.
14. Anderson, R.A. (1974) Formation of bacterial flagellar bundle. Abstr. Symp. Swimming Flying Nat., Calif. Inst. Technol. p. 6.
15. Silverman, M. and Simon M.I. (1972) Flagellar assumbly mutants in *Escherichia coli*. J. Bacteriol. 112, 986—993.
16. Silverman, M. and Simon, M.I. (1974) Flagellar rotation and the mechanism of bacterial motility. Nature 249, 73—74.
17. Weibull, C. (1950) Electrophoretic and titrimetric measurement on bacterial flagella. Acta. Chem. Scand. 4, 260—267.
18. McDonough, M.W. (1965) Amino-acid composition of antigenically distinct *Salmonella* flagellar proteins. J. Mol. Biol. 12, 342—355.
19. Ada, G.L., Nossal, G.J.V., Pye, J. and Abbot, A. (1963) Behaviour of active bacterial antigens during the induction of the immune response. Nature 199, 1257—1262.
20. Asakura, S., Eguchi, G. and Iino, T. (1964) Reconstitution of bacterial flagella in vitro. J. Mol. Biol. 10, 42—56.
21. Leifson, E. (1960) in An atlas of bacterial flagellation. Academic Press, New York.
22. Iino, T. (1962) Curly flagellar mutants in *Salmonella*. J. Gen. Microbiol. 27, 167—175.
23. Iino, T. and Mitani, M. (1967) A mutant of *Salmonella* possessing straight flagella. J. Gen. Microbiol. 49, 81—88.
24. Pietschman, K. (1942) Uber die begeibelung der bakterien. Arch. Mikrobiol. 12, 377—472.
25. Leifson, E., Carhart, C.R. and MacDonald, F. (1955) Morphological characteristics of flagella of *Proteus* and related bacteria. J. Bacteriol. 69, 73—82.
26. Martinez, R.J., Ichiki, A.T., Lundh, N.P. and Tronich, S.R. (1968) A single amino acid substitution responsible for altered flagellar morphology. J. Mol. Biol. 34, 559—564.
27. Asakura, S. and Iino, T. (1972) Polymorphism of *Salmonella* flagella as investigated by means of in vitro copolymerization of flagellins derived from various strains. J. Mol. Biol. 64, 251—268.
28. Astbury, W.T., Beighton, E. and Weibull, C. (1955) The structure of bacterial flagella. Symp. Soc. Exp. Biol. 91, 282—305.
29. Champness, J.N. and Lowy, J. (1967) in Symposium on fibrous proteins. (Crewther, W.G., ed.) p. 106, Butterworth, Australia.
30. Kerridge, D., Horne, R.W. and Glauert, A.M. (1962) Structural components of flagella from *Salmonella typhimurium*. J. Mol. Biol. 4, 227—238.
31. Lowy, J. and Hanson, J. (1965) Electron microscopy studies of bacterial flagella. J. Mol. Biol. 11, 293—299.
32. Champness, J.N. (1971) X-ray and optical diffraction studies of bacterial flagella. J. Mol. Biol. 56, 295—310.
33. O'Brien, E.J. and Bennett, P.M. (1972) Structure of straight flagella from a mutant Salmonella. J. Mol. Biol. 70, 133—152.

34. Finch, J.T. and Klug, A. (1972) In The generation of subcellular structures. First John Innes Symposia, Norwich (Markham, R. and Bancroft, J.B., eds.) North Holland, Amsterdam.

35. Bode, W., Engel, J. and Winklmair, D. (1972) A model of bacterial flagella based on small angle X-ray scattering and hydrodynamic data which indicate an elongated shape of the flagellin protomer. Eur. J. Biochem. 26, 313—327.

36. Lowy, J. and Spencer, M. (1968) Structure and function of bacterial flagella. Symp. Soc. Exp. Biol. 22, 215—236.

37. Finch, J.T. (1972) Tilting experiments on bacterial flagella. Abstr. Proc. 5th Eur. Congr. Elec. Microscopy, p. 578. Institute of Physics, London.

38. Erickson, R.O. (1973) The tubular packing of spheres in biological structures. Science 181, 705—716.

39. Routledge, L.M. (1973) Physical Characterization of flagellar filament protein in *Salmonella* strains SJ25 and SJ814. Ph.D. thesis, University of Pennsylvania.

40. Abram. D., Mitchen, J.R., Koffler, H. and Vatter, A.E. (1970) Differentiation within the bacterial flagella and isolation of the proximal hook. J. Bacteriol. 101, 250—261.

41. Asakura, S. (1970) Polymerisation of flagella and polymorphism of flagella. Advan. Biophys. 1, 99.

42. Gerber, B.R., Routledge, L.M. and Takashima, S. (1972) Self assembly of bacterial flagellar protein: dielectric behavior of monomers and polymers. J. Mol. Biol. 71, 317—337.

43. Burge, R.E. and Draper, J.C. (1971) Structure of bacterial flagella from *Salmonella typhimurium*. Effects of hydration and some strains on the equatorial X-ray diffraction patterns of cast films. J. Mol. Biol. 56, 21—34.

44. Sleytr, U.B. and Glauert, A.M. (1973) Evidence for an empty core in a bacterial flagella. Nature 241, 542—543.

45. Asakura, S., Eguchi, G. and Iino, T. (1968) Unidirectional growth of *Salmonella* flagella in vitro. J. Mol. Biol. 35, 227—236.

46. Iino, T., (1969) Polarity of flagellar growth in *Salmonella*. J. Gen. Microbiol. 56, 227—239.

47. Emerson, S.V., Tokuyasa, K. and Simon, M.I. (1970) Bacterial flagella: polarity of elongation. Science 169, 190—192.

48. Suzuki, H. and Iino, T. (1966) Annu. Rep. Natl. Inst. Genet. Japan 17, 121.

49. Kerridge, D. (1972) in The generation of subcellular structures. First John Innes Symp. Norwich (Markham, R. and Bancroft, J.B., eds.), North Holland, Amsterdam.

50. Berg, H.C. and Andersen, R.A. (1973) Bacteria swim by rotating their flagellar filaments. Nature 245, 381—382.

51. Gerber, B.R., Minakata, A. and Kahn, L.D. (1974) Electric birefringence of bacterial flagellar protein filaments: Evidence for field-induced interactions. In preparation.

52. Iino, T., Suzuki, H. and Yamaguchi, S. (1972) Reconstitution of *Salmonella* flagella attached to cell bodies. Nature New Biol. 237, 238—240.

53. Bütschli, O. (1883) in Klassen und Ordnungen des Thierreichs. (Bronn, H.G., ed.), p. 851, C.F. Winter, Leipzig.

54. Reichert, K. (1909) Ueber die Sichtbarmachung der Geisseln und die Geisselbewegung der Bakterien. Z. Bacteriol. I. Orig. 51, 14—44.

55. Moody, M.F. (1973) Sheath of Bacteriophage T4. III. Contraction mechanism defined from partially contracted sheaths. J. Mol. Biol. 80, 613—636.

55a Gerber, B.R. and Noguchi, H. (1967) Volume change associated with the G-F transformation of flagellin. J. Mol. Biol. 26, 197—210.

56. Gerber, B.R. and Robbins, H. (1971) Biophys. J. Soc. Abstr. 11, 102a.

57. Klug, A. (1967) The design of self-assembly system of equal units. In Formation and fate of cell organelles. Symp. Int. Soc. Cell Biol. (Warren, K.B., ed.) p. 613, Academic Press, New York.

58. Stocker, B.A.D. (1956) Bacterial flagella: Morphology, constitution and inheritance. Symp. Soc. Gen. Microbiol. 6, 19—40.
59. Doetsch, R.N. (1966) Some speculations accounting for the movement of bacterial flagella. J. Theor. Biol. 11, 411—417.
60. Mussill, M. and Jarosch, R. (1972) Bacterial flagella rotate and do not contract. Protoplasma. 75, 465—469.
61. Chwang, A.T. and Wu, T.Y. (1971) A note on the helical movement of micro-organisms. Proc. Roy Soc. London Ser. B 178, 327—346.
62. Shimada, K., Yoshida, T. and Asakura, S. (1974) Cinemicrographic analysis of the movement of flagellated bacteria. Bacterial Locomotion in Abstr. Symp. Swimming Flying Nat. Calif. Inst. Technol., p. 5.
63. DiPierro, J.M. and Doetsch, R.N. (1968) Enzymic reversibility of flagellar immobilization. Can. J. Microbiol. 14, 487—489.
64. Calladine, C.R. (1974) Bacteria can swim without rotating flagellar filaments. Nature 249, 385.
65. Larsen, S.H., Adler, J., Gargus, J.J. and Hogg, R.W. (1974) Chemomechanical coupling without ATP, the source of energy for motility and chemotaxis in bacteria. Proc. Natl. Acad. Sci. U.S. 71, 1239—1243.
66. Macnab, R.M. and Koshland, Jr, D.E. (1972) The gradient-sensing mechanism in bacterial chemotaxis. Proc. Natl. Acad. Sci. U.S. 69, 2509—2512.

Comparative Physiology — Functional Aspects of Structural Materials
Eds L. Bolis, H.P. Maddrell and K. Schmidt-Nielsen
© North-Holland Publishing Company — 1975 — Amsterdam

Bioelectric control of ciliary activity

J.E. DE PEYER and H. HUGGEL

*Department of Animal Biology, Physiological Laboratory, University of Geneva,
Geneva (Switzerland)*

INTRODUCTION

Specific kinds of ciliary activity in protozoa depend closely on membrane activity. The most spectacular of these activities is ciliary reversal, which is induced by membrane depolarization. Although the modalities of this depolarization vary in different species, the fundamental phenomenon is constant throughout the ciliate protozoa. Recent observations have shown clearly that depolarization increases the Ca^{2+} conductance, which increases the inclux of Ca^{2+} [1].

The intracellular concentration of Ca^{2+} then rises to a level sufficient to evoke ciliary reversal.

The role played by the membrane in the control of other parameters, such as for instance the frequency of beat is not yet well understood.

Experiments on Triton-X-extracted cilia of *Paramecium* [2] show that they are reactivated by application of MgATP.

In this case the Ca^{2+} affects only the direction of the movement.

The question arises of how the systems for direction control and frequency regulation are related to each other. In studying the living animal Machemer [3] finds evidence for an important relation between the two phenomena, suggesting that a simple "membrane regulated" agent is responsible for the two activites.

CONTROL OF ORIENTATION

Studies by Eckert and Naitoh [4] using intracellular electrodes, show that hyperpolarization as well as depolarization induces an increase in the frequency of ciliary beating, but also, that the depolarization induces a modification of the beat direction. This modification is not an "all or none" response, but the angle by which the "effective stroke" is shifted with respect to the normal direction varies with the intensity of the depolarizing stimulus and can reach a maximum of 180°C.

How does the membrane regulate this graded ciliary response? In experiments in Triton-X-100 extracted models of *Paramecium* Naitoh and Kaneko [5] show that the ciliary apparatus is reactivated in the presence of magnesium and/or ATP. The frequency of beating of such models is a function of magnesium and ATP and beating can be reactivated even in absence of any other ion ($Ca^{2+} < 10^{-9}$ M). On the other hand, the direction of the movement is strongly if not completely dependent on Ca^{2+}.

For concentrations less than 10^{-6} M the frequency of beating is high, with a slight reduction at about 10^{-7} M. At 10^{-6} M the frequency reaches its maximum, but the swimming velocity of the animal is zero, because the direction of the effective stroke is displaced and is now perpendicular to the longitudinal axis, inducing a rotational movement of the animal. For higher values this direction of the effective stroke shows more and more a reversal and the animal swims backward, but the beating frequency decreases until cessation of movements at 10^{-3} M.

This seems to indicate that in living ciliates a reversal induced by depolarization corresponds to an increase in free intracellular Ca^{2+}.

If it is assumed that the mechanism of ciliary reversal in this case is similar to that of the model, intracellular Ca^{2+} should be about 10^{-7} M, with an external Ca^{2+} concentration about 10^{-3} M. This signifies the existence of a high extra-intracellular gradient.

The leakage current of Ca^{2+} is determined by $I_{Ca} = G_{Ca}(V_m - E_{Ca})$, where I_{Ca} = Ca current; G_{Ca} = Ca conductance; V_m = membrane potential; E_{Ca} = equilibrium potential for Ca.

Consequently, if the depolarization induces a ciliary reversal we suppose that this is caused by an increase in Ca^{2+} membrane conductance, which then induces a net Ca^{2+}-flux down its electrochemical gradient from outside to inside the cell. When entering the cell, Ca^{2+} evoke, because of their positive charges, a depolarization which becomes regeneratively augmented since more Ca^{2+} enters the cell with more depolarization (Ca^{2+} response).

In *Paramecium*, this Ca^{2+} response has a graded character [6], while in other cases (*Stylonychia mytilus*, a hypotrich ciliate) it shows an "all or none" response. However, in the two cases the reversal occurs only if the stimuli are sufficiently strong to induce a regenerative depolarization. The Ca^{2+} dependence of the regenerative response and its relation to the reversal movement is evidenced by the following facts:

(1) In the absence of external Ca^{2+}, both disappear.

(2) The slope of the peak depolarizations, tested in the presence of various cationic concentrations, shows a linear relation to a logarithmic increase of the external Ca^{2+} concentration, close to the ideal value of 29 mV, "predicted from the Nernst-equation for a 10-fold increase in external concentra-

tion of divalent cations''. The effect of other cations, except for Ba^{2+} and Sr^{2+} is much smaller, indicating that the depolarizing membrane current is largely carried by Ca^{2+} [1].

(3) With increased electric stimulus intensities the latency of ciliary reversal drops to 4 ms or less, then increases to a value at which ciliary reversal is delayed until after the end of the stimulation pulse. Duration of reversal increases to several seconds and the corresponding regenerative response increases in amplitude and rate of rise [7].

(4) When pulses shifted the membrane potential to +70 mV, which is assumed to approach the range of the Ca^{2+} equilibrium potential, ciliary reversal was suppressed until the end of the pulse [7].

(5) Failure of the membrane to produce regenerative responses to depolarization will uncouple the response of the ciliary apparatus from the membrane potential, since depolarization under those circumstances will not produce an increased Ca^{2+} influx. In fact, a mutant of *Paramecium aurelia*, named *Pawn* was unable to give ciliary reversal even to extreme depolarizing stimuli [8].

Electrophysiological examination of this mutant showed that depolarization produces only a passive (electrotonic) shift in membrane potential. On the other hand, Triton-extracted models of this mutant swim backwards in the presence of the same Ca^{2+} concentration which produces backward swimming in models of the wild type. This indicates that the inexcitability of *Pawn* is due to an impairment of the mechanism of increased Ca^{2+} conductance in the depolarized state of the membrane; Ca^{2+} does not enter the cell in sufficient amounts to produce ciliary reversal.

Thus, the Ca^{2+} current hypothesis states that reversal of ciliary beating occurs when Ca^{2+} influx produces a sufficient increase in the intraciliary concentration of Ca^{2+}. The existence of an inward depolarizing current, responsible for the upstroke of the action potential was shown with the voltage clamp technique [9].

Depolarization evokes a change in membrane conductance which results in a transient inward current that we now know to be carried by Ca^{2+}. The following steady-state outward current is presumed to be largely carried by potassium due to a delayed increase in K^+ conductance of the membrane. The peak value of the initial inward current increases with increasing depolarization to a maximum value, then decreases. This "negative resistance" portion of the current-voltage relation is responsible for the regenerative nature of the calcium response.

The Ca^{2+} current hypothesis can be summarized by Fig. 1. Any depolarizing stimulus (chemical, electrical or mechanical) results in a transiently increased membrane conductance to Ca^{2+}. According to the current equation, this results in an increment in Ca^{2+} influx. The internal Ca^{2+} concentration rises, activating the mechanism of reversal. Subsequent removal of Ca^{2+} by active processes restores the normal beat orientation of the cilia [10].

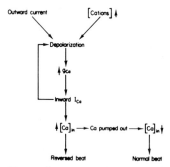

Fig. 1. Sequence of steps leading to ciliary reversal. The calcium current hypothesis. For explanation see text. (After Eckert [10]; with permission.)

Regarding other parameters of the ciliary movement, it seems unlikely that the metachronal wave is dependant on the electronic activity of the cell. [4]. It has been demonstrated in the *Paramecium* models that the beating of the cilia can be reactivated in the absence of an electrically intact membrane. This indicates the existence of intrinsic properties which are independent of bioelectric phenomena.

As to the metachronal coordination of cilia it should be noted that the velocity of the metachronal wave in ciliates is at least 100 times slower than the velocity of signal propagation across the plasma membrane of *Paramecium*. Thus, such signals passively travelling along the plasma membrane can hardly account for the coordination of the cilia. The nearly isopotential condition of the membrane and the passive nature of signal propagation also makes it difficult to conceive in which way the direction of membrane-born impulses might be altered to coordinate changes which occur in the direction of the metachronal wave.

CONTROL OF FREQUENCY

It was already mentionned that a correlation exists between the membrane potential and the frequency of ciliary beating. Depolarization and hyperpolarization both increase the beat frequency. The role of the membrane in controlling this parameter is less well understood. Eckert tentatively assumed that increased frequency of beating in ciliates occurs in response to an increase in the intracellular concentration of Ca^{2+}. This was based on the hypothesis that both hyper- and depolarization cause an increase in Ca^{2+} influx; the first by increase in EMF on Ca^{2+}, the second by a decrease in membrane Ca^{2+} conductance. This hypothesis is supported by the observation by Naitoh [11] that with hyperpolarization a ciliary reversal is obtained in *Opalina* (a parasitic ciliate in the rectum of amphibia) which indicates a certain increase in free intracellular Ca^{2+}.

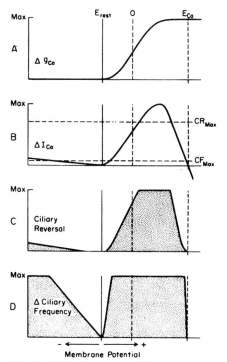

Fig. 2. Eckert's proposed relations between membrane potential, calcium conductance (A), Ca^{2+} current (B), ciliary reversal (C) and ciliary frequency (D). (A) Peak transient increases in Ca^{2+} conductance, Δg_{Ca}, produced by sudden changes in V_m from resting potential (E_{rest}) to various hyperpolarized (toward left) or depolarized (toward right) membrane potentials. (B) Corresponding transient change in inward Ca^{2+} current ΔI_{Ca}. When ΔI_{Ca} is above the level labeled CF_{max}, ciliary frequency is maximal and when ΔI_{Ca} is above the level labeled CR_{max}, ciliary reversal is maximal. (C) Reversed beating increases with the intracellular concentration of Ca^{2+}, which is closely related to ΔI_{Ca} of plot B. (D) Frequency of beating more sensitive to I_{Ca} than is reversal. (After Eckert [10] with permission).

Hence, we may summarize the proposed relation between membrane potential, Ca^{2+} conductance, Ca^{2+} current, ciliary reversal and ciliary beating frequency. A typical S-shaped curve for Ca^{2+} conductance is characteristic of an excitable membrane. Increased Ca^{2+} conductance induces an increase in inward Ca^{2+} current. This in turn, provokes an increase in ciliary reversal and in beating frequency (Fig. 2) [10].

In a recent study, Machemer [3] tries to find a more specific answer to the question of frequency regulation. He shows that the ciliary beating frequency in the unstimulated cell is between 10 and 20 Hz, and the powerstroke toward the posterior right of the cell. Stimulation by hyperpolarization current accelerates beat frequency, which may reach 40—50 Hz. In the same time the ciliary powerstroke is turned clockwise in a more posterior

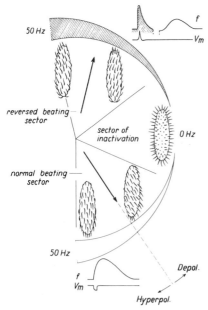

Fig. 3. Machemers proposed relation between directional and frequency responses of the cilia in *Paramecium*. Three sectors of ciliary orientation: (1) Normal beating sector with the power stroke directed toward the posterior and the right (arrow). (2) Reversed beating sector with the power stroke pointing in the anterior and right direction (arrow). (3) Inactivation sector where the directional component is absent. Observed frequency time courses (upper and lower insets) reflect progress of directional and frequency coupling with time. Shaded areas: frequency during reversal. *f*. Frequency time course. V_m, time-course of membrane potential. (After Machemer [3].)

direction. The post-stimulatory decrease in frequency follows a linear slope and is graded with the stimulus strength. Moreover, with decreasing frequency the direction of powerstroke gradually shifts towards the prestimulatory orientation.

We have already seen that a depolarization induces an increase in frequency coupled with a reversal of the powerstroke, which points towards the anterior of the cell. In this case, the post-stimulatory stage shows an exponential decrease in frequency and simultaneousely a gradual clockwise rotation of the ciliary powerstroke. At the end of the period of reversal beating, the cilia enter a state of inactivation, showing no sign of beating. Normal beating does not follow reversal until after the period of inactivation. During inactivation the cilia extend more or less perpendicularly to the cell surface and perform some irregular movements at low frequency and amplitude. After inactivation, normal beating starts at low frequency and then rises gradually to the normal level. Moreover, it was shown that stimulation at the threshold of ciliary reversal depressed the frequency of normal beating (Fig. 3).

It can be concluded that frequency responses are tightly linked to the orientational responses of the cilia.

Therefore, these observations do not support a concept of direct coupling of frequency responses to certain Ca^{2+} levels, since a decrease instead of predicted increase in frequency occurs under two conditions which are both presumably associated with slightly increased intraciliary levels of Ca^{2+} concentration:

(1) Inactivation following reversal beating.

(2) Depression of normal beating activity after small depolarizations.

The inactivation phenomenon between the reversal and the normal ciliary activity remains unexplained.

In conclusion, the coupling between frequency and directional responses of the cilia strongly implicates Ca^{2+} in controlling both parameters of cilia activity. However, a simple positive correlation between intracellular Ca^{2+} concentration and beat frequency appears less likely. The hypothesis according to which hyperpolarization-induced ciliary reactions result from increased inward leakage of Ca^{2+} driven by increased EMF at essentially constant Ca^{2+} conductance, as suggested by Eckert, has still to be substantiated by more stringent data. In fact, the observed effects may also be produced by a decrease of net Ca^{2+} influx through reduced membrane Ca^{2+} conductance. In that case, both increased as well as decreased intracellular calcium concentrations result in an increase of ciliary frequency, the first in the reversed direction and the second in the normal direction.

REFERENCES

1. Naitoh, Y., Eckert, R. and Friedman, K. (1972) A regenerative calcium response in *Paramecium*. J. Exp. Biol. 56, 667—681.
2. Naitoh, Y. and Kaneko, H. (1972) ATP-Mg-reactivated Triton-extracted models of *Paramecium*: modification of ciliary movement by Ca^{2+}. Science 176, 523—254.
3. Machemer, H. (1974) Frequency and directional responses of cilia to membrane potential changes in *Paramecium*. J. Comp. Physiol. 92, 293—316.
4. Eckert, R. and Naitoh, Y. (1970) Passive electrical properties of *Paramecium* and problems of ciliary co-ordination. J. Gen. Physiol. 55, 467—483.
5. Naitoh, Y. and Kaneko, H. (1973) Control of ciliary activities by adenosine triphosphate and divalent cations in Triton-extracted models of *Paramecium caudatum*. J. Exp. Biol. 58, 657—676.
6. Eckert, R. and Naitoh, Y. (1969) Graded calcium spikes in *Paramecium*. Abstr. 3rd Int. Biophys. Congr. Int. Union Pure Applied Biophys., Cambridge, Mass., p. 257.
7. Machemer, H. and Eckert, R. (1973) Electrophysiological control of reversed ciliary beating in *Paramecium*. J. Gen. Physiol. 61, 572—587.
8. Kung, C. and Eckert, R. (1972) Genetic modification of electrical properties in an excitable membrane. Proc. Natl. Acad. Sci. U.S. 69, 93—97.
9. Naitoh, Y. and Eckert, R. (1974) The control of ciliary activity in Protozoa. In Cilia and Flagella, (Sleigh, M.A., ed.) Academic Press, New York and London.

10. Eckert, R. (1972) Bioelectric control of ciliary activity. Science 176, 473—481.
11. Naitoh, Y. (1958) Direct current stimulation of *Opalina* with intracellular microelectrode. Annot. Zool. Jap. 31, 59—73.

Comparative Physiology — Functional Aspects of Structural Materials
Eds L. Bolis, H.P. Maddrell and K. Schmidt-Nielsen
© North-Holland Publishing Company — 1975 — Amsterdam

Principles of contraction in the spasmoneme of verticellids. A new contractile system.

TORKEL WEIS-FOGH

Department of Zoology, University of Cambridge, Cambridge CB2 3EJ (U.K.)

INTRODUCTION

The problem which the present and the succeeding article by Amos [1] attempt to answer is whether we are dealing with a contractile system which is new in principle and not just a variation or a modification of an already known mechanism. There are three basically different types of motile systems known at present. (1) The mechanism responsible for the movements of bacterial flagella. As discussed by Routledge [2], there is growing evidence that the flagellum is actively rotated at its base by some mechanism not yet understood, but it is certain that the main length of the flagellum is built from a single type of protein, flagellin, which is not endowed with any enzymatic properties. (2) The extremely widespread and varied motile mechanisms based on an active sliding between filaments concurrently with the hydrolysis of ATP or related substances. The enzymatic sites may be an integral part of the protein from which the filaments are assembled (cf. muscle myosins) or be part of molecules attached to the filaments (cf. ciliary and flagellar dyneins). Motility by means of actively sliding filaments is not only characteristic of all types of muscular contraction (cf. Pepe et al., Pringle and Haselgrove [3—5]) and of the beating of cilia and flagella [6,7] but evidence accumulated over the last decade strongly suggests that, in spite of many variations, this type of mechanism is responsible for amoeboid movements, streaming of cytoplasm in plant cells, transport inside nerve axons, change in shape of thrombocytes and a great many other movements (see recent review by Pollard and Weihing [8]). Without doubt, active shearing between filaments and molecules of this general nature is the dominating mode of motility in eukaryotic cells. (3) Finally, the contractile mechanism of the so-called myonemes of ciliates, best known in the stalked peritrichs *Vorticella*, *Carchesium*, *Epistylis* and *Zoothamnium*, but probably of the same general nature as the contractile fibrils in the holotrich ciliate *Trachelocerca* and the heterotrichs *Stentor* and *Spirostomum*. In the classical literature the contractile fibrils in the cell body are usually referred to as myonemes or muscle fibrils (Muskelfibrillen, cf. [9]) and the stalk fibre as the stalk muscle (Stielmuskel, cf. [9]). However, since it is now quite clear that they are fundamentally different from muscle fibrils it is preferable to use the name "spasmoneme" first suggested by Entz [10].

HISTORY

In some early theories on the molecular mechanism of muscular contraction, the myofibrils were assumed to be relatively simple rubberlike structures which change from one state to another under the influence of metabolites or ions [11,12]. The discovery of sliding filaments naturally disposed of these theories and today they are mostly forgotten. However, some observations and experiments indicated that the spasmoneme in the contractile stalk of *Vorticella* and *Carchesium* behaves differently from other motile systems and more in accordance with the simple idea. Thus, Schmidt [13] observed that the birefringence of the extended myoneme [9] is high and that it is considerably reduced during shortening, even approaching the isotropic state, in contrast to the small change in muscle. Levine [14] found that glycerinated stalks contract when some divalent ions are added and that they extend when EDTA is admitted. Hoffmann-Berling [15] working in Hans H. Weber's laboratory clearly showed that the contraction of the spasmonemes depends on divalent ions and that ATP did not appear to be directly involved. He obtained the greatest effect with Ca^{2+} and Sr^{2+} in small concentrations and also found that the extended or contracted state can be maintained indefinitely at the appropriate free Ca^{2+} concentrations, also in the presence of strong ATPase inhibitors. Using CaEGTA buffers, Amos [16] confirmed and extended these observations considerably and found that after the intracellular membranes are destroyed by means of detergents, the glycerinated *Vorticella* stalks remain contracted as long as the solution contains 10^{-6} M free Ca^{2+} and remain extended in 10^{-8} M; the threshold is $5 \cdot 10^{-7}$ M and is independent of the absence or presence of Mg^{2+} at least up to $2.5 \cdot 10^{-2}$ M. He could not confirm the claims by Townes and Brown [17] that ATP influences the results but these authors did not use properly defined Ca^{2+} buffers.

According to high-speed films of naturally contracting *Vorticella* stalks taken by Jones et al. [18], the coiling of the stalk and the withdrawal of the cell body as a consequence of the contraction of the spasmoneme lasts only 4 ms at room temperature. Their data enabled Amos [16] to calculate the work done to move the cell body against the viscous drag of the water, and therefore also the rate of work done by the spasmoneme against external forces, excluding the extra work done against the resistance of the extracellular stalk sheath itself. He found that the external work could be accounted for by the change in chemical potential of free, intracellular calcium if the concentration was altered from 10^{-8} to 10^{-6} M and if the amount of Ca^{2+} liberated (probably by the metabolically controlled saccular system inside the spasmoneme) corresponds to $1.1 \cdot 10^{-3}$ mol Ca^{2+} per l of intact spasmoneme. The necessary change then corresponds to 10^{-3} mol of Ca^{2+} per kg wet spasmoneme. This is the minimum amount of Ca^{2+} which must be liberated and absorbed during one cycle of operation in a system dependent only on the change in calcium concentration. Since the spasmoneme contains about 20% dry matter, the energetics require a change of $5 \cdot 10^{-3}$ mol Ca^{2+} per kg dry matter, or about 0.2 g Ca per kg dry protein, in order to account for the work done in vivo. We shall compare this theoretical figure with some recent measurements and also compare the speed of contraction and the power output of shortening spasmonemes with those of striated muscles.

Further advance was difficult until we found and isolated a giant spasmoneme in the colonial freshwater vorticellid *Zoothamnium geniculatum* Ayrton [19]. Engelmann [9] observed the birefringence, structure and contraction of the equally large spasmoneme of a closely related and apparently rare species *Z. arbuscula* Ehrenberg more than 100 years ago, but no further attempts seem to have been made to use it for motility studies.

A GIANT SPASMONEME

There are many colonial vorticellids provided with a common stalk which is contractile at first but becomes inert after the first few divisions of the founder individual. However, according to the detailed observations on *Z. geniculatum* by Wesenberg-Lund [20], the many hundreds or thousands of individual "zooids", all of approximately the same size and shape as the cell body of *Vorticella*, retain structural and functional continuity with each other (Fig. 1) and are provided with individual contractile stalks which fuse to form contractile branches which themselves become united into a single contractile stem. Fig. 2 shows a partially contracted colony. The branching, highly refractile "tree" seen inside the grey envelope of cytoplasm is the spasmoneme system. In a large colony, the stem is of almost uniform cross section (about 30–40 μm) and up to 1 mm long. On stimulation the inverse "umbrella" in Fig. 1 is rapidly drawn into a ball and the common stalk is flexed at the "knee" seen in Fig. 2 which is connected with the end of the spasmoneme via an extracellular "tendon" just visible inside the extracellular sheath of the stalk. The contraction also throws the otherwise straight upper part of the stalk into one turn of a helix so that the whole colony is rapidly whipped towards the supporting leaf to which it is attached.

High-speed cinematography in this laboratory (9200 frames/s (Amos,

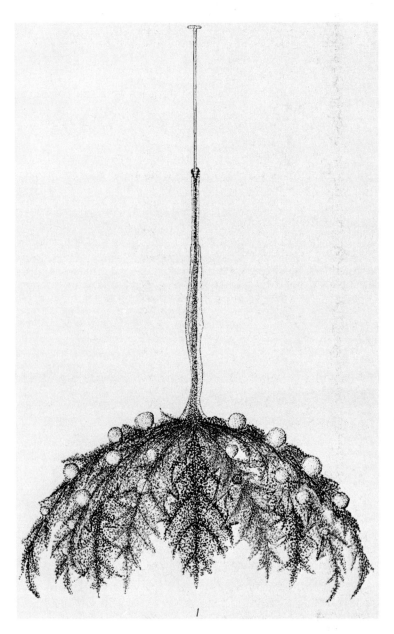

Fig. 1. A fully grown colony of *Z. geniculatum* with its large spherical ciliospores from which new colonies are initiated. The animal is seen as it normally hangs from the underside of floating leaves, drawn by Wesenberg-Lund [20].

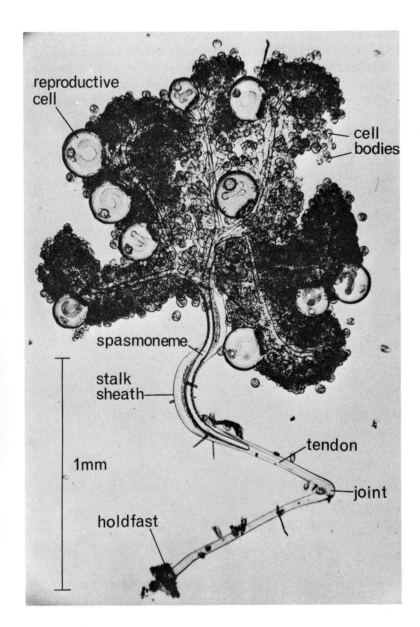

Fig. 2. A partially contracted colony of *Z. geniculatum* after treatment with 2% glutaraldehyde. In the main stalk the common stem of the spasmoneme "tree" is seen as a refractile rod surrounded by the grey envelope of cytoplasm. Note the "knee" and the light extracellular "tendon" which transmits the tensile force of the spasmoneme to the site of insertion at the other side of the "knee".

88

W.B., unpublished)) has shown that (a) the total shortening period lasts 5 ms, (b) the common spasmoneme itself shortens to about 45% of its length L at rest, and (c) the major part of the shortening occurs at constant speed during a period of only 3 ms, so that the shortening speed is about 170 L/s. This is an order of magnitude larger than measured in the fastest striated muscles shortening under zero load [21,22]. It was also easily confirmed that the spasmoneme is strongly birefringent when extended, positive in the longitudinal direction, and that the birefringence decreased drastically with shortening. In conformity with observations on other vorticellids [13,23] the reverse or extension phase is slow and lasts 1—2 s during which the birefringence returns to its original value. Careful analysis of the films showed that the changes in volume, if indeed present to any significant extent, are much too small to account for the length changes.

THE ISOLATED GLYCERINATED SPASMONEME

(For details see Weis-Fogh and Amos [19].)

After glycerination in 50% (v/v) glycerol in a saline buffered at pH 7 and with 2 mM of Na_2EDTA, the stalks were extended and the common spasmoneme could be pulled out from the sheath as a "solid", rubberlike cell organelle of giant dimensions. It was suspended to a specially designed microbalance (Fig. 3) consisting of an ordinary microscope slide provided with a small horizontal platform suspended from two horizontal quartz fibres (Fig. 4). A cover slip resting on spacers completed the small cuvette in which force, extension, diameter and birefringence could be measured simultaneously. The saline solutions used contained either high (10^{-5} M) or low

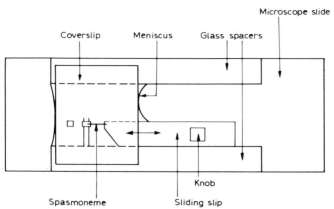

Fig. 3. The isolated giant spasmoneme suspended under a coverslip which rests on spacers. By means of a sliding slip the length and tension of the spasmoneme can be changed manually (from Weis-Fogh and Amos [19]).

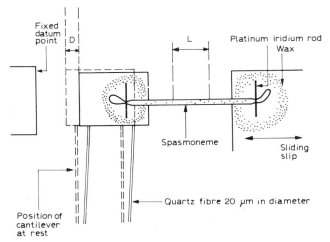

Fig. 4. Details of the cantilever platform and the suspension of the spasmoneme by means of two wax beds into which rods of platinum-iridium are pressed. Grains of graphite on the spasmoneme serve as markers (from Weis-Fogh and Amos [19]).

(10^{-8} M) Ca^{2+} concentrations at pH 7, adjusted by means of CaEGTA buffers. Both at high and at low Ca^{2+} concentrations, Fig. 5 shows that the spasmoneme behaves as a rubberlike body with an elastic modulus G of about $6 \cdot 10^4$ N/m^2. It can be extended reversibly up to 4 times its unstrained length but there is a marked difference between the results obtained in the two solutions. In 10^{-5} M of free Ca^{2+} (circles) the organelle is indis-

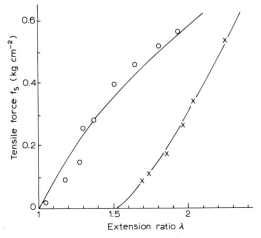

Fig. 5. Force-extension curves for a spasmoneme at high (10^{-5} M; circles) and low calcium ion concentrations (10^{-8} M; crosses). The tension is given as force per unit unstrained cross-sectional area. 1 kg/cm^2 equals $9.81 \cdot 10^4$ N/m^2 (from Weis-Fogh and Amos [19]).

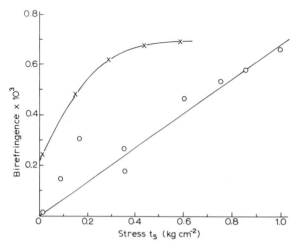

Fig. 6. The birefringence of the same spasmoneme as in Fig. 5 as a function of the stress, i.e. force per unit strained cross-sectional area, at 10^{-5} M Ca^{2+} (circles) and at 10^{-8} M (crosses).

tinguishable from a perfect, lightly vulcanised rubber: the curve drawn is the theoretical relationship between the force f per unit unstrained area and the extension ratio, i.e. $f = G\,v^{1/3}\,(\lambda - \lambda^{-2})$, where $v = 0.15$ is the volume fraction of dry protein in the swollen spasmoneme as estimated by interference microscopy, and λ is the extension ratio and is 1 at zero strain. In accordance with ordinary rubbers [24], the birefringence in Fig. 6 (circles) increases linearly with the stress and is zero when the sample is not strained, i.e. the unstrained spasmoneme is optically and mechanically isotropic.

When the Ca^{2+} concentration was lowered to 10^{-8} M, the unstrained spasmoneme extended spontaneously and could actually exert a push on the platform, a fact which confirms the observation [9] on *Z. arbuscula* and Amos [16] on *Carchesium* that the spasmoneme appears to extend actively during the extension phase in vivo. Also, the unstrained spasmoneme becomes birefringent (Fig. 6, crosses). In other words, the originally isotropic network becomes optically and mechanically anisotropic and now follows a new and steeper force-extension curve which starts at a spasmoneme length more than 50% larger than at 10^{-5} M. The spasmoneme therefore becomes longer and more rigid than before and these changes could not be ascribed to changes in the swollen volume of the organelle.

Superficially, the two curves in Fig. 5 may resemble the length-tension diagrams of resting and contracted muscle, respectively. However, in the case of the spasmoneme they represent true functions of state in the thermodynamic sense and are therefore fundamentally different from the steady-state diagrams from other contractile systems which depend on continuous hydrolysis of ATP or other energy delivering organic molecules. The spasmoneme curves only depend on the maintained level of free Ca^{2+} in a way similar to the unique relationship between temperature, volume and pressure of an ideal enclosed gas. By altering the Ca^{2+} concentration we can therefore bring the spasmoneme round a cycle of states in an anticlockwise direction so that it does work on the surroundings at the expense of the changes in chemical potential of the ion which the experimenter imposes by changing the solution under the cover-slip. This surprisingly simple fundamental principle is reminiscent of the contraction models for muscle proposed by Karrer [11] and later in a more detailed form by Pryor [12]: it may serve to illustrate that even in biology nature sometimes works in the simplest conceivable way.

Ca^{2+} BINDING AND CONTRACTION

Having established the general principle, we must now return to the molecular mechanism. As described by Amos [1,25] the spasmoneme of vorticellids is provided with a dense branched system of membrane-bound saccules which is never more than 100 nm away from the contractile material and which appears to sequester Ca^{2+} [26]. If the saccule system functions like the sarcoplasmic reticulum of striated muscle, we have both a source from which Ca^{2+} can be released as a consequence of stimulation and reabsorbed into during extension, and a source of energy in the form of ATP supplied by the mitochondria. It is significant that destruction of the function of mitochondria and saccules by means of detergents (digitonin and Tween 80) did not affect the response of the glycerinated spasmoneme to Ca^{2+} [19] so that the essential mechanism must reside in the remaining system of longitudinally arranged thin filaments of only 2 nm diameter [16]. 40—60% of the total protein mass of the spasmoneme in *Zoothamnium* consists of a group of closely related proteins from which the filaments appear to be made and which have a molecular weight of about 20 000 (see succeeding article). The problem is whether these proteins react with Ca^{2+} in a specific way and to a sufficient extent to explain the contraction in vivo.

To study this we have recently determined the absolute amount of Ca^{2+} bound by single glycerinated *Zoothamnium* spasmonemes in the extended and contracted states respectively, using X-ray microprobe analysis [27]. The isolated spasmoneme was placed in a drop of the appropriate solution on an aluminium-coated Nylon film, the surplus fluid was removed, the sample dried and, finally, the preparation received a top coating of aluminium in order to increase thermal and electrical conductivity. The stage of the X-ray microprobe analyser (JEOL, JXA-50A) was kept well below —120°C throughout. Fig. 7 shows the transmission electron image at high contrast (too high to show the actual outlines) together with four scanning traces: the straight line is both the line of scanning and the base line for the

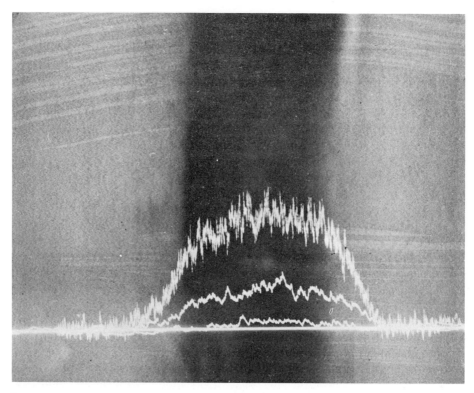

Fig. 7. X-ray signals from a spasmoneme (vertically orientated) when scanned by the electron beam in the transverse direction along the straight line, as explained in the text (from Routledge et al. [27]).

following three count rates, as taken from top to bottom, (a) the total mass present in the direction of the scan, measured by an energy dispersive silicon detector, (b) the signal obtained by a diffracting spectrometer set at peak K_α-radiation for Ca, and (c) the off-peak signal of Ca. The difference between (b) and (c) represents the true amount of Ca present in the spasmoneme. It closely follows the mass curve. The quantitive results are summarised in Fig. 8 and were obtained by scanning rectangular areas in the middle of the spasmoneme. The total amount of Ca^{2+} in the EGTA buffers was varied but the difference in free Ca^{2+} between the contracted and extended states was maintained constant, at 10^{-6} and 10^{-8} M, respectively. In this way we could compensate for accretion effects and show that 1.7 g calcium is bound during contraction per kg of dry spasmoneme protein. Since 40—60% of the protein is the low-molecular weight fraction (20 000), this means that on contraction, 1.4—2.1 calcium atoms are bound per protein molecule, probably two. In other words, the binding is very specific and does not resemble general electrostatic affinity. Also, the affinity for Ca^{2+} is

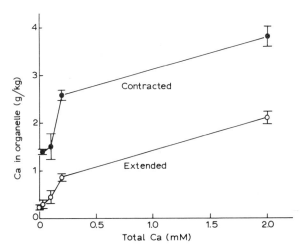

Fig. 8. The total Ca^{2+} content of contracted and extended spasmonemes and its variation with the total Ca^{2+} concentration in the EGTA buffers used, the free Ca^{2+} concentration being 10^{-6} (filled circles) and 10^{-8} M (open circles) throughout the entire range (from Routledge et al. [27]).

extremely high. Even in a Ca^{2+}-free solution containing $2 \cdot 10^{-2}$ M EGTA at pH 7, a small amount of Ca^{2+} is still retained, as seen in Fig. 8.

The amount of Ca^{2+} bound during contraction and released during extension, 1.7 g/kg dry spasmoneme, compares well with the minimum change of 0.2 g/kg needed to drive the contraction cycle in vivo, showing that the proposed system could account for the viscous work within a safety factor of 8. In view of the extreme speed with which spasmonemes contract, this theoretical efficiency seems plausible.

COMPARISON BETWEEN THE PERFORMANCE OF SPASMONEMES AND MUSCLES

Structurally and chemically the two systems are completely different (see [1]). As to performance, Table I summarizes the available information and our deductions.

The fastest known skeletal muscle is the extensor digitorum longus from the mouse. According to Close [21,22] its maximum intrinsic speed when shortening against zero load at $35-37°C$ is 60.5 μm/s per sarcomere which, for a sarcomere length of 2.8 μm, is 22 muscle lengths per second (L/s). Its maximum isometric force P_0 per unit cross sectional area of the myofibrils is not greater than $3 \cdot 10^5$ N/m^2. For optimum power output in skeletal muscle, the load is about one third of P_o and the corresponding speed, one-fourth of the intrinsic speed, or 5.5 L/s, giving an instantaneous power output per m^3 of myofibril of $5.5 \times 1 \times 10^5$ W/m^3, or 550 W/kg wet myofibril. How do these figures compare with those from spasmonemes?

TABLE I

Comparison Between the Performance of Fast Skeletal Muscles of Mammals and of Spasmonemes. See text for details.

	Shortening speed (L/s)	Approx. tensile force (N/m^2)	Instantaneous power output (W/kg wet)
Mouse finger muscle (21,22)			
Intrinsic speed	22	0	0
Speed at maximum power	5.5	10^5	550
Spasmonemes:			
Vorticella in vivo (15)	—	10^4	Approx. 2800
Carchesium, in vivo:			
from Ueda (33) and			
Amos (15)	Av. 120		—
from Rahat et al. (23)	—	$4 \cdot 10^4 - 8 \cdot 10^4$	—
Zoothamnium (19)	172 (in vivo)	$3 \cdot 10^4$ (in vitro)	Approx. 4000

According to Amos [16] the *Vorticella* spasmoneme produces work at a rate of 2750 W/kg against the viscous drag of the water when the cell body is pulled towards the support as a consequence of the coiling of the stalk. The exact in vivo figure for the power output is not known because the spasmoneme also does work against the forces in the stalk and because some elastic energy is stored in the extended spasmoneme. However, the figure must be of the correct order of magnitude. In the case of *Zoothamnium* we may combine the in vivo speed of contraction with the force-extension diagrams measured in vitro and illustrated in Fig. 5. When the fibre shortens from $\lambda = 2$ to $\lambda = 1$, the tensile force varies from about $6 \cdot 10^4$ N/m^2 to zero and the work output is about $3 \cdot 10^4$ J/m^3 or 30 J/kg wet spasmoneme. However, the area under the lower curve corresponds to an amount of elastically stored energy of 10 J/kg so that the net work produced by the spasmoneme as a consequence of the release of Ca^{2+} is 20 J/kg. Since the total contraction time in vivo is 5 ms, the true power output during a normal contraction is likely to be $20/(5 \cdot 10^{-3}) = 4000$ W/kg wet spasmoneme, or of the same order of magnitude as found in the living *Vorticella*. This estimate also tends to support our belief that the isolated glycerinated preparation does represent the essential contractile system fairly accurately. The remaining figures in Table I derive from additional sources. Taken as a whole, there is no doubt that the spasmoneme contracts very much faster than does any striated muscle. In spite of the somewhat lower load carried by spasmonemes as compared with muscle, the instantaneous power output is at least 5 times larger per unit mass than in muscle.

TENTATIVE INTERPRETATION

The giant cilliospores of *Zoothamnium* seen in Figs 1 and 2 become detached from the colony, swarm away and settle to form new colonies. Shortly after attaching itself, the ciliospore forms a new common stalk consisting of a tubular extracellular sheath and a finger-like cytoplasmic extension with the growing giant spasmoneme clearly visible as a refractile rod [20]. The stalk remains extended if undisturbed but the growing, straight spasmoneme is birefringent and contractile when stimulated [9]. This suggests that the spasmoneme is assembled from its 20 000-mol.wt precursors at a low intracellular Ca^{2+} concentration and one must assume that the protein molecules are anisodiametric and polarized so that they can form longitudinal, extended filaments. Since both the extended and the contracted spasmoneme is rubberlike, the oriented molecules must be firmly linked together, although not by means of covalent bonds [1], and the peptide backbone must retain a very considerable kinetic freedom when solvated in water. In a system of this nature, how can we envisage unidirectional contraction promoted specifically by the binding of Ca^{2+}?

Two related but different types of mechanism should be briefly discussed on the basis of the very limited evidence available at present. In the first case we may consider the properties of a polypeptide gel with many fixed negative charges, for instance carboxyl groups, such as analysed by Katchalsky [28]. The interplay between electrostatic repulsion and thermal randomisation or coiling makes it possible to convert chemical energy into mechanical work when the fixed charges are neutralised so that the system collapses into a random thermal network. In the case of carboxyl groups, this could be done by means of H^+ but in our case one must postulate specific electrostatic interaction between the negative groups and Ca^{2+}. However, whatever the detailed mechanism such an electrostatic system is influenced in a nonspecific way by the presence of other ions which tend to shield the fixed charges. Increasing the concentration of neutral salts should then decrease the response to Ca^{2+}. We have no evidence for this at present but more work is needed before we can discard this simple model.

The other model consists of a longitudinal array of polarized molecules such as described at the beginning of this section, each molecule having one or two parts carrying specific Ca^{2+} binding sites while the major part of the chain takes up a random thermally agitated configuration. When Ca^{2+} are bound to the specific sites, the molecules and the network are most random and the configurational entropy highest. When Ca^{2+} are removed, part of the chains become non-random, for instance due to the formation of backbone hydrogen bonds, and the system extends and its entropy decreases.

These two models are not mutually exclusive and may exist simultaneously because the 20 000-mol.wt protein is quite acidic [1]. However, the latter model is more attractive because of the high specific affinity for Ca^{2+} of both the glycerinated spasmoneme (Fig. 8) and the 20 000-mol.wt protein fraction (Amos, this volume). Indeed the low threshold concentration of $5 \cdot 10^{-7}$ M Ca^{2+} for contraction of the *Vorticella* stalk [16] and the fact that some Ca^{2+} cannot be removed from the giant spasmoneme by means of EGTA at pH 7 indicate that, with respect to Ca^{2+} affinity, the main protein(s) of the spasmoneme resembles two well known Ca^{2+}-binding proteins: (a) troponin C from vertebrate muscle [29] (high-affinity binding constant $5.4 \cdot 10^{6}$ mol^{-1}) and (b) the parvalbumins isolated from the muscles of fishes and amphibia. In parvalbumin from hake and frog, Benzoana et al. [30] found binding constants as high as $0.25 \cdot 10^{7}-1 \cdot 10^{7}$ mol^{-1}.

The molecular structure of carp parvalbumin is particularly interesting in this context because Kretsinger and Nockolds [31] have recently shown that both binding sites in the molecule coordinate Ca^{2+} by 6 oxygen atoms in an octahedral arrangement, i.e. a stero-specific and not a general electrostatic binding. They also noted that the binding of Ca^{2+} may confer instability on the protein since 21 main chain hydrogen bonds were not taken up, although this instability appeared to be compensated for by the formation of a large hydrocarbon core. In other words, binding or release of Ca^{2+} appear to cause substantial configurational changes.

On the basis of this new information about Ca^{2+}-binding proteins, it seems more likely than before that the second model may come closer to the truth. With respect to the high speed of shortening of the spasmoneme and therefore its power output, the serial arrangement of molecular "contractile" sites naturally explains these findings, as discussed by Amos [1]. It may be argued, however, that a swollen protein network has too high an internal damping to permit such high rates. Against such arguments stand the results obtained with another rubberlike protein, resilin from insect cuticle. According to Jensen and Weis-Fogh [32], resilin cuticle subject to sinusoidal oscillations of 200 Hz and with a mean strain rate of 40 L/s dissipates less than 10% of its stored energy per cycle as heat and this fraction only increases very slowly with frequency above 100 Hz. There is therefore direct evidence from another rubberlike protein system that high elastic efficiency can be obtained at strain rates comparable to those observed in living spasmonemes.

ACKNOWLEDGEMENTS

The experimental work is the result of a collaborative effort in the Department of Zoology, Cambridge, supported by the Science Research Council. I thank Dr W.B. Amos and Dr L.M. Routledge for many elucidating discussions and, together with Drs B.L. Gupta and T.A. Hall thank them for permission to quote unpublished results. The microprobe work was supported by a grant from the Science Research Council to Prof. T. Weis—Fogh and Drs. P. Echlin, B.L. Gupta, T.A. Hall and R.B. Mouton.

REFERENCES

1. Amos, W.B. (1975) This volume, pp. 99—104.
2. Routledge, L.H. (1975) This volume, pp. 61—73.
3. Pepe, F.A., Chowrashi, P.K. and Wachsberger, P.R. (1975) This volume, pp. 105—119.
4. Pringle, J.W.S. (1975) This volume, pp. 139—152.
5. Haselgrove, J.C. (1975) This volume, pp. 127—138.
6. Summers, K.E. and Gibbons, I.R. (1971) Adenosine triphosphate induced sliding of tubules in trypsin-treated flagella of sea-urchin sperm. Proc. Nat. Adac. Sci. U.S. 68, 3092—3096.
7. Gibbons. B.H. and Gibbons, I.R. (1972) Flagellar movement and adenosine triphosphatase activity in sea-urchin sperm extracted with Triton X-100. J. Cell Biol. 54, 75—97.
8. Pollard, T.D. and Weihing, R.R. (1974) Actin and myosin and cell movement. CRC Critical Reviews of Biochemistry, 2, 1—65.
9. Engelmann, T.W. (1875) Contractilität und Doppelbrechung. Pflüger's Arch. 11, 432—464.
10. Entz, G. (1892) Die elastischen und contractilen Elemente der Vorticellinen. Math. Naturw. Berichte Ungarn. 10, 1—48.
11. Karrer, E (1933) Kinetic theory of muscular contraction. Protoplasma 18, 475—489.
12. Pryor, M.G.M. (1950) Mechanical properties of fibres and muscles. Progr. Biophys. 1, 216—268.
13. Schmidt, W.J. (1940) Die Doppelbrechung des Stieles von *Carchesium* inbesondere die optische-negative Schwankung seines Myonems bei der Kontraktion. Protoplasma 35, 1—14.
14. Levine, L. (1956) Contractility of glycosinated Vorticellidae. Biol. Bull. 111, 319.
15. Hoffman-Berling, H. (1958) Der Mechanismus eines neuen von der Muskelkontraktion verschiedenen Kontraktionszyklus. Biochim. Biophys. Acta, 27, 247—255.
16. Amos, W.B. (1971) Reversible mechanical cycle in the contraction of *Vorticella*. Nature 229, 127—128.
17. Townes, M.M. and Brown, D.E.S. (1965) The involvement of pH, adenosine triphosphate, calcium and magnesium in the contraction of the glycerinated stalks of *Vorticella*. J. Cell Comp. Physiol. 65, 261—270.
18. Jones, A.R., Jahn, T.L. and Fonseca, J.R. (1970) Contraction of protoplasm. IV. Cinematographic analysis of the contraction of some peritrichs. Cell Physiol. 75, 9—20.
19. Weis-Fogh, T. and Amos, W.B. (1972) Evidence for a new mechanism of cell motility. Nature 236, 301—304.
20. Wesenberg-Lund, C. (1925) Contributions to the biology of *Zoothamnium geniculatum* Ayrton. Kgl. Danske Vidensk. Selsk. Skrifter Nat. Math. Afd., Raekke X, 1, 1—53.
21. Close, R. (1965) The relation between intrinsic speed of shortening and duration of the active state of muscle. J. Physiol London 180, 542—559.
22. Close, R.I. (1972) Dynamic properties of mammalian skeletal muscles. Physiol. Rev. 52, 129—197.
23. Rahat, M., Pri-Paz, Y. and Parnas, I. (1973) Properties of stalk-'muscle' contractions of *Carchesium* sp. J. Exp. Biol. 58, 463—471.
24. Treloar, L.R.G. (1958) The physics of rubber elasticity, pp. 1—342, Clarendon Press, Oxford.
25. Amos, W.B. (1972) Structure and coiling of the stalk in the peritrich ciliates *Vorticella* and *Carchesium*. J. Cell Sci. 10, 95—122.
26. Favard, P. and Carasso, N. (1965) Mise en évidence d'un réticulum endoplasmique dans le spasmonème de ciliés péritriches. J. Microsc. 4, 567—572.

27. Routledge, L.M., Amos, W.B., Gupta, B.L., Hall, T.A. and Weis-Fogh, T. (1975) Microprobe measurements of calcium binding in the contractile spasmoneme of a vorticellid. J. Cell Sci. in the press.
28. Katchalsky, A. (1964) Polyelectrolytes and their biological interactions. In Suppl. to Biophys. J. 4: Connective tissue intercellular macromolecules, 9—41.
29. Hartshorne, D.J. and Pyun, H.Y. (1971) Calcium binding by the troponin complex, and the purification and properties of troponin A. Biochim. Biophys. Acta 229, 698—711.
30. Benzoana, G., Capony, J.P. and Pechere, J.F. (1972) The binding of calcium to muscular parvalbumins. Biochim. Biophys. Acta 278, 110—116.
31. Kretsinger, R.H. and Nockolds, C.E. (1973) Carp muscle calcium-binding protein. II. Structure determination and general description. J. Biol. Chem. 248, 3313—3326.
32. Jensen, M. and Weis-Fogh, T. (1962) Biology and physics of locust flight. V. Strength and elasticity of locust cuticle. Philos. Trans. Ser. B 245, 137—169.
33. Ueda, K. (1954) Electrical stimulation of the stalk muscle of *Carchesium*. II. Dobutsugaku Zasshi. 63, 1—14 (mainly in Japanese).

Comparative Physiology — Functional Aspects of Structural Materials
Eds L. Bolis, H.P. Maddrell and K. Schmidt-Nielsen
© North-Holland Publishing Company — 1975 — Amsterdam

Structure and protein composition of the spasmoneme

W.B. AMOS

Department of Zoology, University of Cambridge, Cambridge CB2 3EJ (U.K.)

The general construction of a vorticellid ciliate is that the cell body is attached to some solid support, such as a submerged leaf, by means of a stalk. The outer casing and much of the interior of the stalk is secreted material but there is within this casing an extension of the cytoplasm containing the spasmoneme. The organelle is so named because it brings about the rapid flexure of the stalk when it contracts, the form of bending being determined chiefly by the pattern of reinforcing fibres in the extracellular casing of the stalk [1,2]. The spasmoneme is continuous with other contractile fibres, or myonemes, in the cell body and it does not differ from them in fine structure. This is interesting, since it can contract independently of the myonemes [3].

Because of its high refractive index, the spasmonene can be seen readily with the light microscope. It lies within a strand of optically granular cytoplasm, in which the granules are spherical mitochondria. Especially in *Carchesium* and *Zoothamnium* the spasmoneme may appear cleft or frayed locally into longitudinal strands, suggesting a fibrous substructure.

This is borne out by electron microscopy [2,4,5]. So far, only contracted material has been examined because the organisms invariably contract strongly when placed in fixatives. The principal component visible in sections of the organelle is a dense mass of filaments, each with a diameter of 2—3 nm, chiefly longitudinal in orientation. No membrane separating the filamentous mass from the remainder of the stalk cytoplasm has been found, and it seems likely that there is no special resistance to diffusion between the two zones, since mitochondria occur embedded in the interior of the mass in *Zoothamnium* as well as in the surrounding cytoplasm.

Intact filaments cannot easily be separated from the isolated spasmoneme of *Zoothamnium*, but small fragments of filamentous material were obtained by soaking glycerinated organelles briefly in distilled water before transferring them to saturated aqueous uranyl acetate as negative stain. These frag-

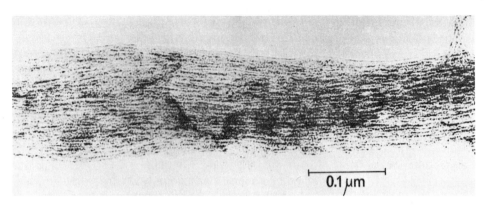

0.1 μm

Fig. 1. Fragment of a glycerinated spasmoneme of *Zoothamnium* negatively stained with saturated aqueous uranyl acetate.

ments appear to consist of bundles of filaments of similar size to those seen in thin sections. The negatively stained filaments have an indistinct beaded appearance (see Fig. 1) with a longitudinal periodicity of about 3.5 nm.

The spasmoneme also contains a system of membranous sacs in the form of interconnected longitudinal tubules. These tubules are at least several micrometers long and may possibly extend throughout the entire length of the stalk though their ultimate fate in the cell body is unknown. The walls of the tubules sometimes, though rarely, present a trilaminar appearance like that of other cell membranes. The usual diameter of the tubules is 38—70 nm but they may be dilated more than 10 times in some regions, probably as a result of poor fixation. Carasso and Favard [6] have shown cytochemically that the tubules are capable of accumulating Ca^{2+} in their lumina, and have advanced the idea that they function like the sarcoplasmic reticulum of striated muscle in controlling the Ca^{2+} concentration around the contractile apparatus. The high rate of activation of the spasmoneme (3 ms in a *Zoothamnium* spasmoneme 30 μm in diameter) is in accord with activation by release of Ca^{2+} from sites throughout the contractile mass, such as would occur if the walls of the tubules suddenly became permeable to Ca^{2+}. However, it is intriguing in view of this rapid activation that no structural counterpart of a T-system is visible in the spasmoneme; none of the tubules runs transversely or makes contact with the plasma membrane. Allen [5] has described intricate fibrillar connections between the myonemes and plasma membrane which may perhaps serve some conducting function.

The spasmonemes of *Vorticella*, *Carchesium* and *Zoothamnium* are similar in fine structure, the chief differences being in gross dimensions. *Vorticella*, which is a solitary organism, has a spasmoneme with a diameter of 1 μm, while that of *Carchesium* may be 10 μm in diameter. In *Zoothamnium geniculatum* incomplete separation after cell division leads to the formation of a colony of several thousand individuals, all of which remain connected to one branched spasmoneme, the main trunk of which may be more than 1 mm long and 30—40 μm in diameter. This relatively enormous organelle offers an opportunity for chemical analysis unrivalled by any other myonemal structure in a protozoan [7]. Not only does the contractile organelle form a high proportion of the total dry mass of the colony, but also it can be mechanically isolated from the cell by direct dissection. The dry mass of the spasmonemal material in a single mature *Zoothamnium* colony is about 0.5 μg.

The organelle is insoluble in KCl solutions but can be dissolved totally in 1% sodium dodecylsulphate and partially in solutions of 8 M urea and guanidine · HCl. By means of a specially designed apparatus the proteins from one or a small number of spasmonemes were separated by discontinuous electrophoresis in a polyacrylamide slab gel. After glycerination as described by Weis-Fogh and Amos [8], spasmonemes were dissected out in the glycerol medium, rinsed in 50% glycerol to remove salts and then dissolved in a solution containing 2% sodium dodecylsulphate, 0.06 M Tris buffer (pH 6.8), and 0.04% fluoroscein as tracking dye. The organelles dissolved within 30 s in this solution leaving no visible residue. After electrophoresis in 15% acrylamide gels according to the formula of Laemmli and Favre [9] the proteins were stained in Coomassie Blue or Fast Green.

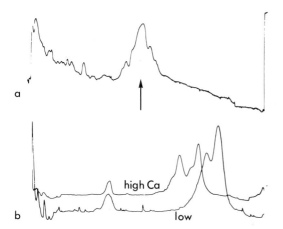

Fig. 2. Densitometer tracings of polyacrylamide gels containing spasmoneme proteins stained with Coomassie Blue. The direction of migration during electrophoresis was to the right (anode). (a) Sodium dodecylsulphate-extracted proteins from 10 spasmonemes separated in a 15% acrylamide sodium dodecylsulphate gel. The 20 000 mol. wt peak is indicated. (b) The effect of free Ca^{2+} concentration on the electrophoretic mobility of spasmoneme proteins. By means of EGTA the high Ca^{2+} gel was buffered at 10^{-6} M Ca^{2+}, the low at 10^{-8} M Ca^{2+}. Stacking gel composition: 3% acrylamide, 0.08% bis-acrylamide, 10^{-3} M $CaCl_2$, $2.5 \cdot 10^{-2}$ M Tris/HCl (pH 8.0). Running buffer: $4 \cdot 10^{-2}$ M Tris/glycine (pH 8.0), $5 \cdot 10^{-4}$ M $CaCl_2$. The running buffer was also adjusted to pCa 6 or pCa 8 as appropriate by means of EGTA.

The sodium dodecylsulphate-gel pattern (Fig. 2a) contained a prominent band corresponding to a molecular weight of 20 000. Most of the remaining material ran slowly, indicating molecular weights above 100 000. The gels were scanned directly with a densitometer and the 20 000 mol. wt band was found to contain 60% of the stain, the two stains giving the same result. Peaks at the molecular weight of actin and tubulin were not found and densitometry showed that these proteins, if present, must comprise less than 2% of the total stainable material. These results show the fundamental dissimilarity between the spasmoneme and other motile structures in cells.

Since sodium dodecylsulphate may denature proteins, other means of dissolving the spasmonemes were tested in an attempt to obtain native protein. The spasmoneme was found to dissolve almost completely in 3 M guanidine · HCl leaving a ghost from which no protein could be extracted with sodium dodecylsulphate. The guanine · HCl extract was freed of salt by dialysis and shown by electrophoresis of the dialysate to contain all the principal protein components of the organelle. These proteins could also be obtained without the labour of dissection by first removing the cell bodies from *Zoothamnium* colonies by a homogenization procedure and then extracting the denuded stalks in 3 M guanidine · HCl. The material with a molecular weight of 20 000 was examined on non-sodium dodecylsulphate-gels and by isoelectric focussing, and found to be quite acidic, with an isoelectric point of 4.7—4.8.

Proteins extracted with guanidine · HCl were used as the starting material for the following series of experiments, designed to test the effect of divalent ions on the electrophoretic mobility of the proteins.

After salt had been removed from the guanidine · HCl extract dialysis was continued against 20 mM EDTA and then against distilled water to remove divalent ions. Electrophoresis in polyacrylamide was performed at pH 8 without the addition of sodium dodecylsulphate (at this pH the glycerinated spasmoneme still contracts in response to Ca^{2+}). In one series of experiments the free Ca^{2+} in the gel was adjusted to 10^{-8} M by means of EGTA, the total Ca^{2+} level being $1 \cdot 10^{-3}$ M. During electrophoresis the 20 000 mol. wt material formed a complex leading band consisting of a prominent peak with a trailing shoulder (see Fig. 2b). In another Ca^{2+}-buffered gel, also containing $1 \cdot 10^{-3}$ M total Ca^{2+}, but with the free CA^{2+} level increased to 10^{-6} M, the leading band was markedly retarded relative to the other spasmoneme proteins, which did not change their mobility. Also, the leading band became more obviously divided into two principal components at the higher Ca^{2+} level. These effects were not observed with Mg^{2+} at the same concentrations, though a decrease in mobility occurred with $5 \cdot 10^{-4}$ M Mg^{2+}.

It is concluded from the electrophoresis results that neither actin nor tubulin forms an appreciable proportion of the mass of the spasmoneme, but there is, however, a characteristic acid protein, or possibly a class of similar acid proteins with molecular weights close to 20 000. This type of protein appears to bind Ca^{2+} with a high specificity in the same range of Ca^{2+} concentrations as induce contraction in glycerinated spasmonemes [10]. The doubling of the retarded peaks in Fig. 2b may perhaps indicate the formation of two kinds of Ca^{2+} complex, one with a single Ca^{2+} bound and the other with two. The observed changes in relative mobility indicate a specific alteration in charge or conformation of the protein, and it is likely that this has a bearing on the mechanism of contraction.

Proof that this protein is present in the filaments visible in the electron microscope is at present lacking but the beaded appearance of the filaments could be due to the presence of subunits with a molecular weight of 20 000. Assuming that each filament consists of protein molecules of a single type connected to form a chain by non-covalent bonds, it may be imagined that upon excitation, one or a small number of Ca^{2+} combine with each molecule. This could induce a change in molecular shape or in the manner of bonding of each molecule to its neighbours in such a way as to bring about a shortening of the filament. A mechanism such as this involves the summation in series of many microscopic shortening events and could explain the high intrinsic shortening velocity of the spasmoneme.

An important unsolved problem is the relationship between the proposed subunit construction of the spasmoneme and its rubberlike optical and mechanical properties [8]. It is not known whether the long-range extensibility of the spasmoneme discussed in the previous paper involves an extensibility of the subunits themselves or, for instance, a change in the relative orientation of the bonds between subunits. When Ca^{2+} is removed from the fully contracted and optically isotropic material an anisotropic extension occurs and the spasmoneme becomes birefringent. Theories of the molecular structure of the spasmoneme must therefore incorporate directional properties in the low-Ca^{2+} state as well as ideal rubber-like properties when Ca^{2+} is bound. Further structural and biochemical analysis will be needed to solve these problems.

ACKNOWLEDGEMENTS

The chemical part of this work was supported by an SRC grant to Professor T. Weis-Fogh, whom I thank for encouragement and advice. I also thank Dr L.M. Routledge for permission to quote unpublished results on this aspect.

REFERENCES

1. Fauré-Fremiet, E. (1905) La structure de l'appareil fixateur chez les Vorticellidae. Arch. Protistenk. 6, 207—226.

2. Amos, W.B. (1972) Structure and coiling of the stalk in the peritrich ciliates *Vorticella* and *Carchesium*. J. Cell Sci 10, 95—122.
3. Jones, A.R., Jahn, T.L. and Fonseca, J.R. (1970) Contraction of protoplasm. IV. Cinematographic analysis of the contraction in some peritrichs. J. Cell Physiol. 75, 9—20.
4. Favard, P. and Carasso, N. (1965) Mise en évidence d'un reticulum endoplasmique dans le spasmonème de ciliés péritriches. J. Microscopie 4, 567—572.
5. Allen, Richard D. (1973) Structures linking the myonemes, endoplasmic reticulum and surface membranes in the contractile ciliate *Vorticella*. J. Cell Biol. 56, 559—579.
6. Carasso, N. and Favard, P. (1966) Mise en évidence du calcium dans les myonèmes pédonculaires de ciliés péritriches. J. Microscopie 5, 759—770.
7. Amos, W.B, Routledge, L.M. and Yew, F.F. (1975), submitted for publication.
8. Weis-Fogh, T. and Amos, W.B. (1972) Evidence for a new mechanism of cell motility. Nature 236, 301—304.
9. Laemmli, U.K. and Favre, M. (1973) Maturation of the head of bacteriophage T4. I. DNA packaging events. J. Mol. Biol. 80, 575—599.
10. Amos, W.B. (1971) A reversible mechanochemical cycle in the contraction of *Vorticella*. Nature 229, 127—128.

Comparative Physiology — Functional Aspects of Structural Materials
Eds L. Bolis, H.P. Maddrell and K. Schmidt-Nielsen
© North-Holland Publishing Company — 1975 — Amsterdam

Myosin filaments of skeletal and uterine muscle

FRANK A. PEPE, PROKASH K. CHOWRASHI and PHYLLIS R. WACHSBERGER

Department of Anatomy, Medical School, University of Pennsylvania, Philadelphia, Pa. 19174 (U.S.A.)

SUMMARY

The observations made on thick sections in electron microscopy and on the interaction of LMM* and C-protein in LMM aggregates strongly support the model for the detailed packing of myosin molecules in the myosin filament which was derived by relating antibody staining in the A band and the structural characteristics of the A-band and myosin filaments [4,5]. These observations are not compatible with the models recently proposed by Squire [9]. In addition to the observations discussed in this presentation the model [4,5] predicts the detailed structure of the M-band, since in deriving the model the relation between the M-band and the structure of the myosin filament was considered. The predicted structure of the M-band has been observed [18] and it is incompatible with the model for the M-band proposed by Knappeis and Carlsen [21].

Although there are differences between skeletal and uterine myosin filaments as deduced from studies of the growth of synthetic skeletal [2] and uterine [1] myosin filaments, it is most likely that the packing of the LMM in the backbone of the filaments is the same; and that the difference in the filaments is primarily in the relative orientation of the myosin molecules.

INTRODUCTION

The ultimate objective of these studies is to make a comparative study of the detailed packing of myosin molecules in skeletal and uterine myosin filaments. Most studies have been done on skeletal myosin filaments. Recently it has been possible to obtain highly purified uterine myosin [1]. Using this, we have been able to compare the formation of synthetic uterine myosin filaments with previous observations of the formation of synthetic skeletal myosin filaments [2].

---

* LMM, light meromyosin

Myosin is the major protein component of the myosin filament. The myosin molecule is essentially a rod with a globular region at one end [3]. A portion of the rod, the LMM portion, carries the solubility characteristics of the myosin molecule, i.e. solubility at high ionic strength and insolubility at low ionic strength. The globular region and the rest of the rod are soluble even at low ionic strength. The globular region carries the ATPase and actin combining properties of the myosin molecule and is a part of the myosin cross bridge.

Huxley [2] showed that synthetic skeletal myosin filaments could be grown from solutions of myosin by decreasing the ionic strength of the solution. The synthetic filaments and natural filaments were structurally similar except that the synthetic filaments varied in length while natural filaments were all approx. 1.6 μm in length. From these studies it was concluded that the first aggregation of myosin molecules occurred tail-to-tail, i.e. with the LMM portions overlapping and the globular heads sticking out at each end. Further aggregation occurred in a head to tail fashion at each end with the tails forming the backbone of the filament and the heads on the surface available as myosin cross bridges. Therefore the myosin molecules in one half of the filament are oriented opposite to those in the other half of the filament. In the small region in the middle of the filament where tail-to-tail overlap occurs there are no myosin cross bridges.

The first model for the detailed packing of myosin molecules in the skeletal myosin filament was derived by relating the antimyosin staining pattern observed in electron microscopy to the other structural features of the A-band and of individual myosin filaments [4,5]. The staining pattern observed consisted of seven lines in the middle of each half of the A-band (Fig. 1). Recently it has been shown that the antimyosin prepared against conventionally purified myosin contains antibody specific for C-protein [6,7] (Offer, G. and Pepe, F.A., in preparation) which is a protein contaminant of myosin preparations [8]. It has also been shown that C-protein is associated with the myosin filament [6,7] (Pepe, F.A., Craig, R., Offer, G. and Drucker, B., in preparation) and that the seven lines previously observed with antimyosin staining in electron microscopy [5] are due to specific staining of C-protein.

In deriving the detailed model for the myosin filament [4,5] the spacing between these lines was measured as 43 nm and, it was assumed that this represented the repeat periodicity of myosin molecules along the entire filament; the fact that the lines were not present along the entire length of the filament was related to possible differences in the lateral packing of the LMM along the backbone of the filament [5]. Later I will describe some studies of the interaction between LMM and C-protein in LMM aggregates, which strengthen the validity of these assumptions.

Other possibilities for the detailed packing of the myosin in the myosin filament have recently been proposed by Squire [9]. Where appropriate I will relate these to the model proposed earlier [4,5].

Fig. 1. Antibody staining. Chicken pectoralis muscle stained with antiserum to conventionally purified myosin. The lines are due to specific staining of C-protein [7].

I will now give a general description of the model [4,5] describe some observations which are related to the model and then describe the studies of the interaction of LMM and C-protein in LMM aggregates and how this is related to the model (Chowrashi, P.K. and Pepe, F.A. in preparation). Finally I will describe some studies made on the formation of synthetic myosin filaments from solutions of uterine myosin [1] and relate these to the formation of synthetic filaments from solutions of skeletal myosin.

MODEL FOR THE MYOSIN FILAMENT

The detailed model [4,5] is made up of twelve subunits where each subunit represents linearly aligned myosin molecules staggered by 86 nm so that the HMM portion of one molecule overlaps the LMM portion of the next molecule in the linear aggregate. The twelve linear aggregates are hexagonally packed in parallel as shown in the diagrammatic representation of a cross

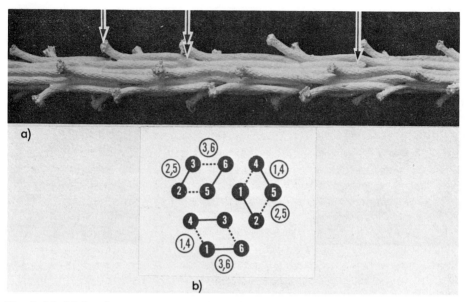

Fig. 2. Model for myosin filament. (a) Scale model constructed using 3.7 nm center to center distance between structural units (one or two myosin molecules [15]). Single arrows indicate the ends of a myosin molecule or structural unit. Double arrow indicates where the myosin molecule overlaps the next myosin molecule in a linear aggregate. (b) Diagrammatic representation of a cross section through the model. Solid circles represent packing of the LMM portion of the molecules in the backbone of the filament. The open circles represent the position of the HMM portion of the molecules excluded to the surface of the filament. A difference of one in the numbers represents a stagger of 14.3 nm. The two numbers in the open circles differing by 3 indicate that the HMM portions of myosin molecules staggered by 43 nm are linearly positioned along the surface of the filament.

section through the model in Fig. 2b. The solid circles represent the packing of the LMM part of the linear aggregates in the backbone of the filament and the open circles represent the soluble HMM portions which are excluded to the surface of the filament. The numbers indicate the relative stagger between the linear aggregates and therefore between the myosin molecules in different linear aggregates. A difference of one indicates a stagger of 14.3 nm. Note that the linear aggregates are arranged in three sets of four. Within a set of four there are two pairs. Within a pair the stagger is 43 nm (numbers differ by 3) and between pairs the stagger is 14.3 nm. The stagger of 43 nm within a pair is consistent with the finding that myosin dimers with a head-to-tail stagger of 43 nm exist in solutions of myosin [10—14]. Linear aggregation of these dimers staggered by 86 nm would produce the two linear aggregates of a pair in the model. The two numbers in each open circle indicate the positions of the HMM portions excluded to the surface, the HMM portions from two linear aggregates staggered by 43 nm being linearly aligned on the surface of the filament. Note that in each set of four linear aggregates the soluble HMM portions from one of them must project between the LMM portions of two surface molecules to be excluded from the LMM packing region or backbone of the filament. This means that the myosin molecules in each group of four linear aggregates are not all equivalent. In Fig. 2a is a picture of this model built to scale using 3.7 nm as the center-to-center distance between the linear aggregates. This figure came from measurement of the center-to-center distance between subunits observed in cross sections of the myosin filament [15] which will be discussed later. With this center-to-center distance it is possible to have a linear aggregate consisting of pairs of myosin molecules instead of single myosin molecules linearly aligned. In the model there are two myosin cross bridges every 14.3 nm and if we have pairs of myosin molecules instead of single myosin molecules linearly aligned we would have two myosin molecules per cross bridge or four myosin molecules every 14.3 nm along the filament. This is consistent with the recent findings of Morimoto and Harrington [16] that there are four myosin molecules for every 14.3 nm along the myosin filament. In Fig. 2a the two single arrows point out the ends of a myosin molecule (or pair of molecules). The double arrow is at the junction between the LMM and HMM portions of the myosin molecule. Half way between the double arrow and the farthest single arrow, the HMM portion of a myosin molecule comes to the surface between the LMM portions of two surface molecules, the head end of the molecule ending just behind the double arrow.

This model predicts that in cross section through the myosin filament it should be possible to see twelve linear aggregates. Using thin section in electron microscopy, on rare occasions, clearly defined and well organized subunits can be seen [15]. Examples are shown in Fig. 3. Note that the subunit organization is enhanced on rotation printing of the image, but that the rotated image corresponds closely to the unrotated image. This correspondence is necessary to eliminate the possibility of artifacts being introduced by the rotation [17]. The subunit structure predicted by the model can be clearly seen in Fig. 3. Since such well organized structure is so rarely seen in thin sections we have recently been using thick sections observed in a 200 kV JEM electron microscope. If the model is correct and the linear aggregates are aligned in parallel, a larger amount of material will be directly superimposed in a thick section and the subunits should be more easily visualized. Also since the LMM backbone of the filament is the most rigid structurally the triangular cross sectional profile of the backbone should become more easily visible throughout the length of the filament. In thin sections the triangular cross sectional profile is consistently seen only in the small region of the filament without cross bridges [4,5,18]. In all of the models proposed by Squire [9] the myosin molecules are tilted relative to the long axis of the filament in such a way that in thick sections any

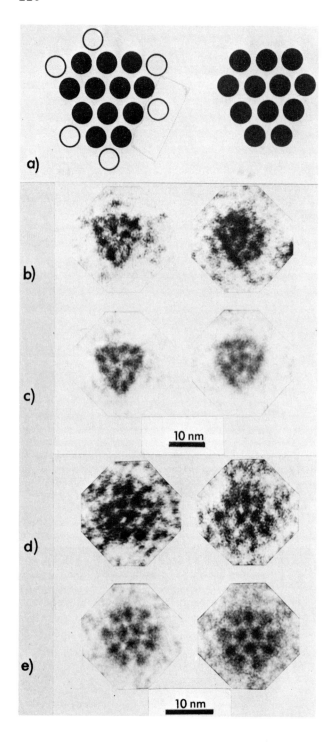

a)

b)

c)

10 nm

d)

e)

10 nm

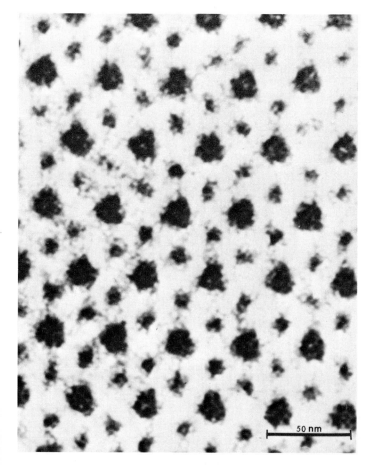

Fig. 4. Cross section of fish muscle. This section is approx. 400 nm in thickness and was observed using a 200 kV (JEM) electron microscope. The cross section is taken through a portion of the filaments where myosin cross bridges are present. Note clear triangular profiles and clear indication of subunit structure.

Fig. 3. Subunit structure of the filament [15]. (a) Diagrammatic representation of the subunit structure of the filament with and without the position of the HMM portions of the myosin molecules indicated. Solid circles represent the LMM portions of the twelve linear aggregates of molecules in the model. Open circles represent the HMM portions of the molecules excluded to the surface. (b) Cross sections of myosin filaments from the pectoralis muscle of the chicken. These cross sections were taken close to the middle of the filament where triangular cross sectional profiles are clearly seen. (c) Rotation printing of the images in (b) using 1/3 of the total exposure time for each of three superimposed exposures differing by rotation of the image through 120°C. (d) Cross section of a myosin filament from the pectoralis muscle of the chicken. These cross sections were taken in a region of the filament where there are myosin cross bridges. Both images are of the same filament. (e) Rotation printing of the images in (d) as described in (c). The two images represent two different rotation printings of the same filament. Note 12 subunits hexagonally packed.

indications of subunit structure would be obliterated and the cross sectional profiles would be more clearly circular. In Fig. 4 is a cross section through myosin filaments in the regions where there are myosin cross bridges. This section is approx. 400 nm in thickness. Note the clarity of the triangular cross sectional profiles. Also note that the filaments are made up of clearly defined subunits. The problem that remains is to find the proper conditions of fixation and embedding which will uniformly preserve the organization of the subunits. These observations however unequivocally demonstrate that increasing the section thickness enhances the observation of subunits and enhances visibility of the triangular cross sectional profile of the backbone of the filament. Both of these characteristics are predicted by the model and they effectively exclude the models proposed by Squire [9].

LMM AGGREGATION AND C-PROTEIN INTERACTION

An important consideration in the derivation of the model was the interpretation of the lines observed in the A-band after antibody staining [4,5] and now known to represent specific staining of C-protein [6,7] (Pepe, F.A., Craig, R., Offer, G. and Drucker B., in preparation). As I have already noted, these lines were originally assumed to represent the repeat periodicity of myosin molecules along the entire filament and their presence along only a portion of the myosin filament was assumed to reflect differences in the lateral packing of the LMM along the backbone of the filament [4,5].

We have found that the binding of C-protein to LMM aggregates can be related to the packing of the LMM in the aggregates (Chowrashi, P.K. and Pepe, F.A., in preparation). In Fig. 5 are the sodium dodecylsulphate-polyacrylamide gels of three preparations of LMM prepared by papain digestion of myosin and which have been purified by column chromatography on DEAE-Sephadex [19]. The preparation shown in Fig. 5a is a mixture of LMM with a chain weight of approx. 72 000 and the entire rod fragment of myosin with a chain weight of approx. 110 000. This preparation was further digested with trypsin to convert the rod to LMM thus giving a single LMM component with a chain weight of approx. 72 000 (Fig. 5b). More extensive digestion with papain gave an LMM with two components having chain weights of approx. 72 000 and 52 000 (Fig. 5c). On aggregation the two first preparations (Figs 5a and 5b) gave a 14.3-nm axial repeat periodicity (Figs 6a and 6b) and the third preparation (Fig. 5c) gave two lines every 43 nm, the separation between the lines being 15—16 nm (Fig. 6c). The 14.3-nm axial repeat period observed in the absence of rod (Fig. 5b) was clearly more distinct than that obtained in the presence of rod (Fig. 5a) indicating that the presence of rod interfered with the packing of the LMM. On aggregation of these LMM preparations in the presence of C-protein (Fig. 7) it was found that when rod was present C-protein binding to the

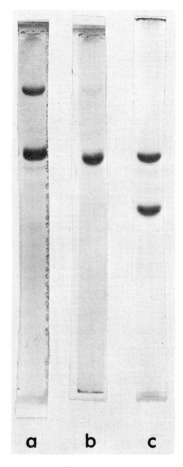

Fig. 5. Sodium dodecylsulphate-polyacrylamide gels of column purified LMM preparations. (a) LMM prepared by papain digestion of myosin. The top band is the rod fragment of myosin with a chain weight of approx. 110 000. The other band is LMM with a chain weight of approx. 72 000. (b) Trypsin digestion of the preparation in (a) was used to eliminate the rod fragment yielding only the 72 000 chain weight LMM. (c) LMM prepared by papain digestion of myosin. The two bands represent two LMM components with chain weights of approx. 72 000 and 52 000.

LMM aggregate occurred at 43 nm and there was no indication of a 14.3-nm axial repeat in the LMM aggregates (Fig. 7a). In the absence of rod there was a clear 14.3-nm axial repeat and no evidence of C-protein binding (Fig. 7b). In the presence of the 52 000 mol. wt LMM component the C-protein was bound at 43 nm intervals (Figs 7c and 7d) and binding was restricted to the 15—16-nm space between the two lines every 43 nm (Fig. 7d). Moos [20] has shown that C-protein can bind to the surface of LMM aggregates which show a 43-nm axial repeat, enhancing the 43-nm repeat periodicity, but the

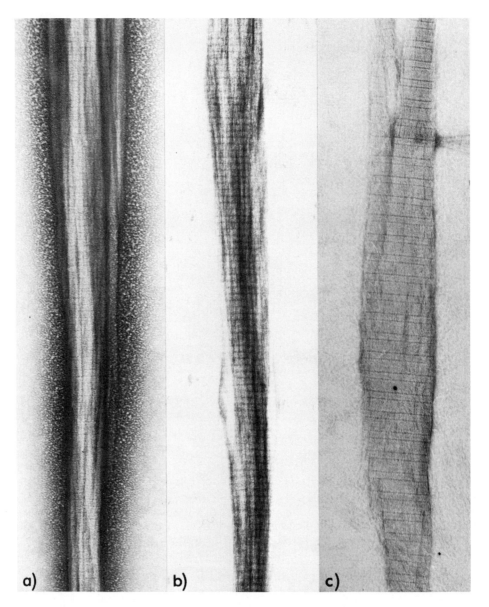

Fig. 6. Aggregation of the LMM preparations. (a) Aggregation of an LMM and rod mixture (as in Fig. 5a) by dialysis into 0.008 M phosphate buffer/0.075 M KCl (pH 7.35). Note weakly defined 14.3-nm axial repeat periodicity. (b) Aggregation of the 72 000 chain weight LMM component (as in Fig. 5b) by dialysis into 0.03 M imidazole buffer/ 0.075 M KCl (pH 7.35). Note clearly defined 14.3-nm axial repeat periodicity. (c) Aggregation of a mixture of the 72 000 and 52 000 chain weight LMM components (as in Fig. 5c) by dialysis into 0.03 M imidazole buffer/0.075 M KCl (pH 7.35). Note two lines every 43 nm. The distance between the two lines is 15—16 nm.

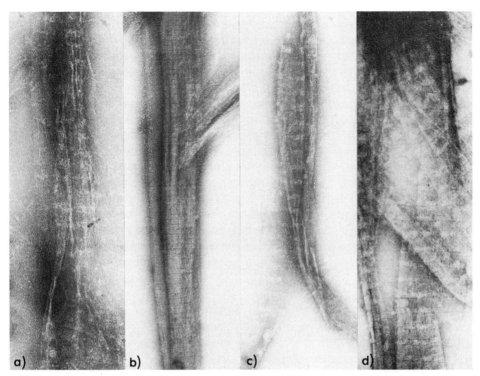

Fig. 7. Aggregation of LMM in the presence of C-protein. (a) Using an LMM and rod mixture (as in Fig. 5a) and the same buffer as in Fig. 6a. (b) Using only the 72 000 chain weight LMM (as in Fig. 5b) and the same buffer as in Fig. 6b. (c) Using a mixture of the 72 000 and 52 000 chain weight components (as in Fig. 5c) and the same buffer as in Fig. 6c.

LMM and the C-protein repeat were not simultaneously observed. Therefore, since there is an error of approx. 2 nm in the measurements made in electron microscopy, it was not possible to conclude unequivocally that the repeat periodicity of LMM and C-protein were identical. In Fig. 7d where both the LMM and C-protein repeats are visible it is possible to conclude that they are identical and that the C-protein repeat is determined by the LMM (or myosin) to which it is bound.

Three important points can be made from the observations presented here: (1) C-protein does not bind when the 72 000 chain weight component is present alone and a 14.3-nm axial repeat is observed. (2) When either rod or the 52 000 chain weight component is present in addition to the 72 000 component, C-protein will bind at 43-nm intervals and will interfere with formation of a 14.3-nm axial repeat. (3) Since the axial repeat pattern of the LMM and the bound C-protein can be seen simultaneously in Fig. 7d and the repeat periodicities are the same, it is unequivocal that the repeat periodicity of the C-protein is identical to that of the LMM to which it is bound. The 72 000 chain weight component would correspond to an LMM which is approx. 86 nm in length. This corresponds to the length of the LMM aggregated in the core of the filament in the model (Fig. 2) [4,5]. The presence of rod would interfere with the tight packing of this LMM because a portion of the rod would overlap other molecules if the packing is in the form of linear aggregates as proposed in the model. This looser packing would permit binding of C-protein to LMM's and rods staggered by 43 nm and the bound C-protein would inhibit interactions of LMM with a stagger of 14.3 nm. The presence of the 52 000 chain weight component which is approx. 64 nm in length would cause gaps to occur in the aggregate likewise resulting in a looser structure and permitting binding of C-protein. The fact that C-protein binding is never superimposed on a 14.3-nm axial repeat suggests that it can only bind to molecules staggered by 43 nm.

These observations and conclusions strongly suggest that: (1) The C-protein periodicity in the myosin filament reflects the repeat periodicity of the myosin to which it is bound. (2) The binding of C-protein to a particular portion of the filament is determined by the lateral packing of the myosin in the backbone of the filament. (3) C-protein can only bind to myosin molecules, in the filament, which are staggered by 43 nm. Therefore the assumptions used to derive the model [4,5] that the lines observed with antibody staining represent the repeat periodicity of myosin along the entire myosin filament and that visibility of the lines along only a portion of the filament reflects differences in lateral aggregation in the backbone of the filament, are substantiated by these studies of the interaction of LMM and C-protein. As has already been mentioned, the myosin molecules in this model are not all equivalent and this lack of equivalence may contribute to the differences in LMM packing in different parts of the filament which determine the position of binding of C-protein.

In the models proposed by Squire [9] the myosin molecules are all equivalent and therefore C-protein binding can only be related to differences in stagger of the myosin molecules in different portions of the filament. Only one of the models proposed by Squire has such differences. In this model there are bridges every 14.3 nm only in the portion of the filament where C-protein binds, with intervals of 28.6 and 43 nm between bridges everywhere else. As we have already seen C-protein binding does not occur when a 14.3-nm stagger predominates but does occur when a 43-nm stagger predominates. Therefore none of the models proposed by Squire [9] is compatible with the results of the studies of LMM and C-protein aggregation (Chowrashi, P.K. and Pepe, F.A., in preparation).

UTERINE MYOSIN FILAMENTS

From studies of the growth of synthetic uterine myosin filaments we have

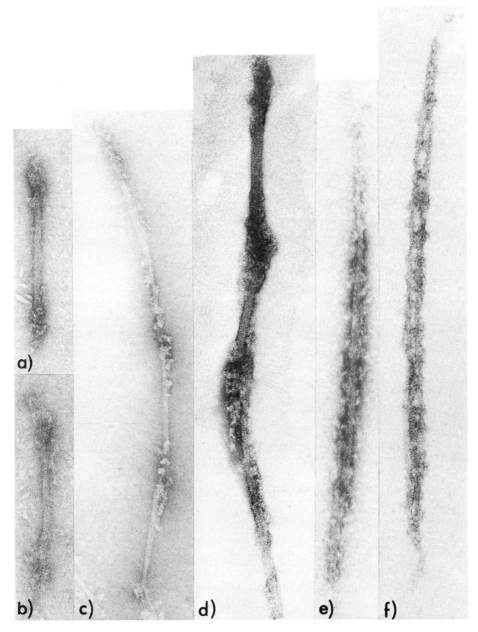

Fig. 8. Formation of synthetic myosin filaments from solutions of column purified uterine muscle myosin. (a) and (b). Short bipolar filaments formed in 0.2 M KCl/0.01 M phosphate buffer (pH 7.0). (c) Linear end-to-end aggregation of short bipolar filaments observed in 0.1 M KCl/0.01 M phosphate buffer (pH 7.0). (d) Linear end-to-end aggregation of short bipolar filaments observed in 0.2 M KCl, 0.01 M phosphate buffer (pH 7.0) containing 10 mM MgCl$_2$. (e) and (f). Long filaments observed in 0.1 M KCl, 0.01 M phosphate buffer (pH 7.0) containing 10 mM MgCl$_2$. Magnification 140 000 ✕.

found that there are differences in the growth of synthetic uterine and skeletal myosin filaments [1]. The first aggregates formed from solutions of uterine myosin on decreasing the ionic strength were essentially identical to those obtained with skeletal muscle myosin. They were bipolar with a smooth portion which corresponds to the region of tail-to-tail (or LMM to LMM) overlap and with the globular ends of the molecule sticking out at both ends. In Fig. 8a and b examples of these bipolar aggregates obtained in 0.2 M KCl can be seen. On decreasing the KCl concentration to 0.1 M, linear end to end aggregation of these bipolar aggregates was also observed (Fig. 8c). Similar experiments were done in the presence of 10 mM $MgCl_2$. In this case the bipolar aggregates were first observed in the presence of 0.3 M KCl and the linear end to end aggregation of these bipolar aggregates was first observed in the presence of 0.2 M KCl (Fig. 8d). Further reduction of the KCl concentration to 0.1 M in the presence of 10 mM $MgCl_2$ resulted in long filaments up to 1200 nm in length with bridges all along the filaments and no evidence of smooth regions without cross bridges (Figs 8e and 8f). In some cases one or more smooth regions could be detected at some place along the long filament suggesting that these represented intermediate stages between the linear aggregates of bipolar filaments and long filaments with bridges along the entire surface of the filament.

From these observations we conclude that: (1) Increase in length of synthetic uterine myosin filaments requires 5—10 mM $MgCl_2$ [1] whereas increase in length of synthetic skeletal myosin filaments does not. (2) The initial aggregation of myosin molecules to form synthetic myosin filaments occurs identically for both uterine and skeletal muscle myosins, i.e. by tail-to-tail overlap to form short bipolar filaments with a smooth middle zone and with bridges sticking out at each end. (3) Further growth of synthetic uterine and skeletal myosin filaments is different. Skeletal myosin filaments grow by head to tail aggregation of myosin molecules at the ends of the short bipolar filaments. Uterine myosin filaments grow by end to end aggregation of the bipolar filaments followed by obliteration of the smooth zones to give a long filament with bridges along the entire length. Therefore, the myosin molecules in long uterine myosin filaments are oppositely oriented all along the filament in contrast to skeletal myosin filaments where all the molecules in one-half of the filament are oriented opposite to those in the other half of the filament.

These differences between uterine and skeletal myosin filaments may or may not reflect differences in the basic organization of the LMM in the backbone of the filaments. It is most likely that this basic packing is identical for the two filaments and that the difference is primarily in the relative orientation of the myosin molecules, i.e. whether neighboring molecules are similarly or oppositely oriented.

ACKNOWLEDGEMENT

This work was supported by U.S.P.H.S. Grant HL 15835 to the Pennsylvania Muscle Institute.

REFERENCES

1. Wachsberger, P.R. and Pepe, F.A. (1974) Purification of uterine myosin and synthetic filament formation. J. Mol. Biol. 88, 385—391.
2. Huxley, H.E. (1963) Electron microscope studies on the structure of natural and synthetic protein filaments from striated muscle. J. Mol. Biol. 7, 281—308.
3 Lowey, S. (1971) In Subunits in Biological Systems, (Timasheff, S.N. and Fasman, G.D., eds), Part A, pp. 201—259, Marcel Dekker, New York.
4. Pepe, F.A. (1966) in Electron Microscopy (Uyeda, R., ed.), Vol. II, pp. 53—54, Maruzen Co., Tokyo.
5. Pepe, F.A. (1967) The myosin filament. I. Structural organization from antibody staining observed in electron microscopy. J. Mol. Biol. 27, 203—225.
6. Offer, G. (1972) C-protein and the periodicity in the thick filaments of vertebrate skeletal muscle. Cold Spring Harbor Symp. Quant. Biol. 37, 87—93.
7. Pepe, F.A. (1972) The myosin filament: immunochemical and ultrastructural approaches to molecular organisation. Cold Spring Harbor Symp. Quant. Biol. 37, 97—108.
8. Starr, R. and Offer, G. (1971) Polypeptide chains of intermediate molecular weight in myosin preparations. FEBS Lett. 15, 40—44.
9. Squire, J. (1973) General model of myosin filament structure. J. Mol. Biol. 77, 291—323.
10. Godfrey, J.E. and Harrington, W.F. (1970). Self-association in the myosin system at high ionic strength. I. Sensitivity of the interaction to pH and ionic environment. Biochemistry 9, 886—893.
11. Godfrey, J.E. and Harrington, W.F. (1970). Self-association in the myosin system at high ionic strength. II. Evidence for the presence of a monomer ⇌ dimer equilibrium. Biochemistry 9, 894—908.
12. Harrington, W.F. and Burke, M. (1972) Geometry of the myosin dimer in high-salt media. I. Association behavior of rod segments from myosin. Biochemistry 11, 1448—1455.
13. Harrington, W.F., Burke, M. and Barton, J.S. (1972) Association of myosin to form contractile systems. Cold Spring Harbor Symp. Quant. Biol. 37, 77—85.
14. Burke, M. and Harrington, W.F. (1972) Geometry of the myosin dimer in high-salt media. II. Hydrodynamic studies on macromodels of myosin and its rod segments. Biochemistry 11, 1456—1462.
15. Pepe, F.A, and Drucker, B. (1972) The myosin filament. IV. Observation of the internal structural arrangements. J. Cell Biol. 52, 255—260.
16. Morimoto, K. and Harrington, W.F. (1974) Substructure of the thick filament of vertebrate striated muscle. J. Mol. Biol. 83, 83—97.
17. Friedman, M.J. (1970) A re-evaluation of the Markham rotation technique using model systems. J. Ultrastructure Res. 32, 226—236.
18. Pepe, F.A. (1971) In Progress in Biophysics and Molecular Biology (Butler, J.A.V. and Noble, D., eds), Vol. 22, pp. 77—96, Pergamon Press, New York.
19. Richards, E.G., Chung, C.S., Menzel, D.B. and Olcott, M.S. (1967) Chromatography of myosin on diethylaminoethyl-Sephadex A-50, Biochemistry 6, 528—540.

20. Moos, C. (1972) Discussion: interaction of C-protein with myosin and light meromyosin. Cold Spring Harbor Symp. Quant. Biol. 37, 93—95.
21. Knappeis, G. and Carlsen, F. (1968) The ultrastructure of the M line in skeletal muscle, J. Cell. Biol. 38, 202—211.

Comparative Physiology — Functional Aspects of Structural Materials
Eds L. Bolis, H.P. Maddrell and K. Schmidt-Nielsen
© North-Holland Publishing Company — 1975 — Amsterdam

Molecular differentiation and myoblast fusion during myogenesis in culture

BIANCA ZANI and MARIO MOLINARO

Institute of Histology and General Embryology, University of Rome, Rome (Italy)

SUMMARY

Myosin synthesis and myoblast fusion in culture appear to be differential-ly regulated. In fact, following inhibition of gene transcription by actinomy-cin D (0.05 μg/ml of growth medium for 8 h), myosin synthesis decreases very slightly during the pre-fusion period (35—43 h of culture) while it is drastically inhibited after the onset of cell fusion. At later periods after fusion (100 h of culture) myosin synthesis is almost completely unaffected by the treatment.

Myoblast fusion is completely insensitive to inhibition of RNA synthesis with actinomycin. These data suggest that myosin synthesis requires simulta-neous gene activity during the period of maximum increment of synthesis while myoblast fusion is supported by stable mRNA.

INTRODUCTION

During myogenesis in culture cell fusion and myosin synthesis seem to be strictly correlated events. In fact a very low level of myosin synthesis is detectable before fusion and several conditions affecting cell fusion impair myosin synthesis as well [1,2]. An interesting problem to be clarified is whether the expression of the two differentiative events is regulated by a common mechanism.

In the experiments described below this problem has been approached by inhibiting gene transcription with actinomycin D. It is shown that in these conditions myoblast fusion is completely unaffected while myosin synthesis is inhibited at different levels during culture. Myosin synthesis occurring during the pre-fusion period is only slightly affected by inhibition of gene transcription, while that occurring immediately after the onset of fusion, when the maximum stimulation of synthesis takes place, is drastically reduced by the antibiotic.

At a later period of growth in culture, the effect of the drug is declining and myosin synthesis in the treated cells is about the same level as in the control.

MATERIALS AND METHODS

Culture

A cell suspension was prepared by trypsin treatment of breast muscle from 10-day-old chick embryos and seeded in collagen coated 9-cm plastic Petri dishes. The conditions of growth were previously described [3,4]. Morphological observations and cell counts were made on plates fixed and stained with Wright's solution.

Actinomycin Treatment and Cell Labelling

The cells were incubated during growth in culture with actinomycin D (0.05 μg/ml of culture medium) for a period of 8 h before harvesting. [^{3}H]-leucine (5 μCi/ml) and [^{3}H]lysine (5 μCi/ml) were added to the medium 1 h after the onset of actinomycin treatment for a total period of 7 h.

Myosin Preparation and Electrophoresis

After incubation with radioactive amino acids the plates were washed with cold saline, and the cell suspension was centrifuged at 800 $\times$ g for 10 min. Myosin was extracted from the pellet after addition of 0.5 $A_{280\ nm}$ carrier myosin prepared from chick embryo as described [6]. Myosin was then dissociated into subunits by incubation with 1% sodium dodecylsulphate and 1% dithiothreitol for 2 h at 37°C and the heavy subunit was isolated by sodium dodecylsulphate-polyacrylamide gel electrophoresis [5].

Counting of Radioactivity

Gels were sliced in sections 2-mm thick. The slices were dried, solubilized in 30% H_2O_2 at 50°C for 2—4 h and counted in a Packard Tri-Carb liquid scintillation spectrometer in a dioxane-containing scintillation mixture.

Incorporation of radioactive leucine and lysine into total protein was measured by precipitating an aliquot of cell homogenate on a Millipore filter and counting the acid precipitable material in a scintillant containing 40 ml Permafluor (Packard) per l of toluene.

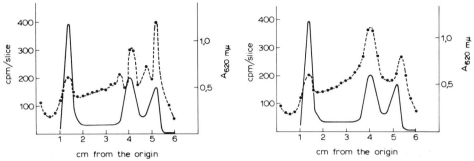

Fig. 1. Electrophoretic pattern of myosin synthesized in culture 43 h after plating. (A) Control. (B) Treated cultures. The continuous line (———) indicates the densitometric trace of carrier myosin after staining the gel [6]. The first peak is myosin heavy subunit, while the peaks with higher mobility are actin, light subunits, and unresolved contaminating proteins. Labeled myosin was prepared from 7 plates in a final volume of 0.480 ml and aliquots of 95 μl were layered on top of the gel in (A) and 140 μl in (B). ■------■ cpm/slice; ———, $A_{620 \, nm}$.

RESULTS

In the conditions of incubation with actinomycin used in these experiments (0.05 μg/ml for 8 h), [^{3}H]uridine incorporation showed about 95% inhibition within 1 h after the onset of treatment, while cell duplication in culture was completely unaffected.

The effect of treatment with the antibiotic on myosin synthesis was measured at different periods of growth in culture. The period of incubation with actinomycin was 8 h and labeled amino acids were added 1 h after

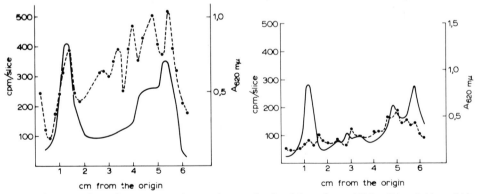

Fig. 2. Electrophoretic pattern of myosin synthesized in culture 76 h after plating. (A) Control. Myosin was prepared from 6 plates in a final volume of 0.250 ml, and an aliquot of 45 μl was layered on top of the gel. (B) Treated cultures. Myosin was prepared from 6 plates in a final volume of 0.450 ml and an aliquot of 25 μl was layered on top of the gel. ——— $A_{620 \, nm}$; ■------■, cpm/slice. Other details are in legend of Fig. 1.

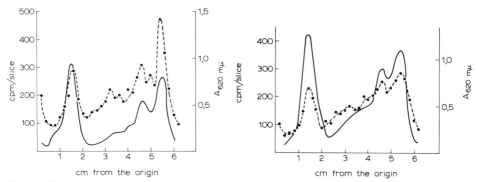

Fig. 3. Electrophoretic pattern of myosin synthesized in culture 100 h after plating. (A) Control. Myosin was prepared from 6 plates in a final volume of 400 μl and 45 μl were layered on top of the gel. (B) Treated cultures. Myosin was prepared for 6 plates in a final volume of 400 μl and 60 μl were layered on top of the gel. ———, $A_{620\,nm}$; ■-----■, cpm/slice. Other details are in legend of Fig. 1.

addition of the antobiotic and were present, therefore, for 7 h. Figs. 1—3 shows the electrophoretic pattern of myosin synthesized in treated and control cultures in the interval between 35 and 100 h growth. These experiments show that the inhibitor has almost no effect on the counts incorporated in the heavy subunit peak at 43 and 100 h, while a drastic reduction of synthesis occurs at 76 h.

The amount of myosin heavy subunit synthesized in culture was calcu-

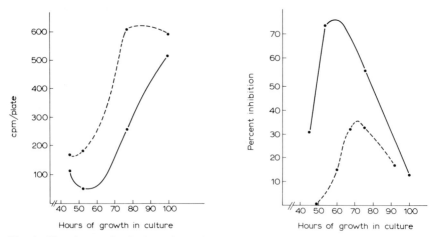

Fig. 4. Myosin synthesis in culture in presence of actinomycin D. ■-----■, control. ■———■, actinomycin D. Each point is an average of three experiments.

Fig. 5. Inhibition of amino acids incorporation into myosin (■———■) and total protein (■-----■) by actinomycin D.

Fig. 6. Fiber formation in vitro in presence of actinomycin D. ■-----■, control; ■——————■, treated cultures. The drug was added to the growth medium 8 h before each experimental point.

lated on the basis of the counts incorporated into the electrophoretic peak and is reported in Fig. 4. The relative inhibition of synthesis after treatment is shown in Fig. 5. The data reported indicate that in the period immediately preceding fusion (35—43 h) a low level of synthesis occurs which is partially resistant to the drug, while in the period following the onset of cell fusion (68—76 h) myosin synthesis is strongly affected by the antibiotic. After this period the inhibitory effect is declining and almost disappears at 100 h. The discontinuous line in Fig. 5 represent the inhibition of amino acid incorporation into total protein. The fact that the incorporation into total protein is very slightly inhibited and that the maximum level of inhibition is reached at a different period of culture than that of myosin synthesis rules out the possibility that ribosome concentration or factors involved in translation could become rate limiting as effect of treatment.

The treatment with actinomycin D in the conditions used is completely ineffective in preventing cell fusion. In fact 8 h incubation with the drug at different periods of growth has no effect on the onset of cell fusion and on the total number of nuclei in syncytia (Fig. 6).

DISCUSSION

The data reported show that myosin synthesis occurring at 35—43 h and 92—100 h of growth in culture does not requires simultaneous gene activity and is therefore presumably supported by previously synthesized mRNA molecules, while in the period between 53 and 76 h of growth in culture, myosin synthesis is largely dependent on continuous gene transcription. The resistance of myosin synthesis occurring in the pre-fusion period (43 h) to the inhibition of transcription could be tentatively explained by assuming the presence in culture of two populations of myogenic cell engaged in myosin biosynthesis in different periods. One population, in a more advanced stage of differentiation, was probably already engaged in myosin synthesis at the onset of culture. This population might be responsible for the early myosin synthesis which is resistant to the actinomycin at 43 h since it is supported by mRNA molecules previously accumulated in the cytoplasm. A second population less differentiated and quantitatively more relevant might be activated during culture to synthesize myosin, between 53 and 76 h. The sensitivity to actinomycin D during this period would therefore depend on an impairment of new gene transcription. An alternative possibility would be the sequential activation of two different genes in the same cell. This interpretation is supported by recent observations showing that two different myosins appear at different moments of myogenesis in the embryo. The absence of an effect of actinomycin on fiber formation in culture supports the possibility that cell fusion is regulated by pre-formed messenger RNA molecules. Our data indicate therefore that the two main differentiative characters of myogenesis, cell fusion and myosin seem to be independently regulated.

REFERENCES

1. Bichoff, R. and Holtzer, H. (1970) Inhibition of myoblast fusion after one round of DNA synthesis in 5-Bromodeoxyuridine. J. Cell Biol. 44, 134—142.
2. Peterson, B. and Stromman, R.C. (1972) Myosin synthesis in cultures of differentiating chicken embryo skeletal muscle. Develop. Biol. 29, 113—138.
3. Molinaro, M. and Martinozzi, M. (1973) Relative contribution of different classes of myogenic cells to muscle fiber formation in culture. Exp. Cell Res. 78, 329—334.
4. O'Neill, M. and Stockdale, F.E. (1972) A kinetic analysis of myogenesis in vitro. J. Cell Biol. 52, 52—65.
5. Dow, J. and Stracher, A. (1971) Changes in the properties of myosin associated with muscle development. Biochemistry 10, 1316—1321.
6. Molinaro, M., Zani, B. Martinozzi, M. and Monesi, V. (1974) Selective effects of actinomycin D on myosin biosynthesis and myoblast fusion during myogenesis in culture. Exp. Cell Res. 88, 402—405.
7. Yaffe, D. and Dym, M. (1972) Gene expression during differentiation of contractile muscle fiber. Symp. Quant. Biol. 37, 543—547.

Comparative Physiology — Functional Aspects of Structural Materials
Eds L. Bolis, H.P. Maddrell and K. Schmidt-Nielsen
© North-Holland Publishing Company — 1975 — Amsterdam

Structural changes in smooth and striated muscle during contraction

J.C. HASELGROVE

MRC Laboratory of Molecular Biology, Hills Road, Cambridge CB2 2QH (U.K.)

INTRODUCTION

Electron microscope and X-ray observations indicate that muscles which are functionally as different as vertebrate striated muscle, vertebrate smooth muscle and invertebrate smooth muscles (indeed all muscle so far studied) all operate by a sliding filament mechanism. The actin-containing filaments in different muscles all have basically the same structure, and similar changes occur when the muscles contract. In contrast to the similarity of the actin structure, the myosin containing filaments vary significantly. I will here describe the structure of three very different muscles, and describe the changes in the structure when the muscles contract.

VERTEBRATE STRIATED MUSCLE

The structure of vertebrate striated muscle is well known (e.g. see review by Huxley [1]). Fibres consist of an axially repeating unit, the sarcomere, in which overlapping thick and thin filaments lie parallel to the fibre axis and arranged in a precise longitudinal and lateral array (Fig. 1). The thick myosin-containing filaments are about 150 Å in diameter and are arranged at the lattice points of a hexagonal lattice. The thin filaments are about 80 Å in diameter and lie at the trigonal positions of the myosin lattice: they contain actin, tropomyosin, and troponin which is sensitive to Ca^{2+} and controls the interaction of actin and myosin [2]. The actin filaments attach to the Z line structure at the end of the sarcomere in such a way that the filaments are polarised in opposite directions on opposite sides of the Z line [3]: the opposite ends of the thick filaments are also polarised in opposite directions so that the force generated between adjacent actin and myosin filaments is always in a direction which will cause the muscle to shorten. Such a polarity of the actin and myosin filaments is a necessary requirement for a sliding filament mechanism which has more than one actin filament and one myosin filament in series.

Myosin Filaments

The heads of the myosin molecules project from the surface of the thick filaments at regular positions in a helical arrangement, and give rise to a series of meridional reflections and layer-lines in the low angle X-ray diffraction pattern [4]. The heads (which are called cross-bridges in structural studies) project at intervals of 143 Å along the filament and give rise to the strong reflection on the meridian at 143 Å, and to the series of layer-lines near the meridian (Figs 2a and 2b) Huxley and Brown [4] studied the diffraction pattern from contracting muscles and found that the layer-line pattern was much weaker than from relaxed muscles, indicating that a large proportion of cross-bridges had moved from their resting position. This finding supported the model of contraction in which cross-bridges attach to actin, move to produce a relative sliding motion between the actin and myosin filaments, and then detach again. A study of the equatorial X-ray reflections from resting and contracting muscle [5] indicated that during contraction the cross-bridges move laterally away from the myosin filaments and closer to the actin filaments than they are in relaxed muscle, and fewer than 50% of the cross-bridges are instantaneously attached to actin. The cross-bridges are still attached to the myosin filament by part of the tail of the molecule (see [1]). A further interesting observation on the X-ray patterns was that the spacing of the two strongest meridional reflections increased by about 1% when the muscle contracts, indicating that the myosin filaments are about 1% longer in a contracting muscle than in a relaxed muscle [4,6].

It was important to see if the cross-bridges that moved during contraction did so only when they were attached to actin. Haselgrove [6] studied muscles which were stretched until only half of the cross-bridges could interact with actin, and Huxley [7] studied muscles stretched so that the actin and myosin filaments no longer overlapped and interaction of the cross-bridges with actin should have been prevented. They concluded that during contraction the movement of the cross-bridges (seen as a decrease in intensity of the X-ray layer-line pattern (Figs 2b and 2c) and the slight 1% increase in length of the filaments is not due to the interaction of each cross-bridge with actin.

The study of muscles in rigor is useful in the investigation of the mechanism of contraction because it is thought that the permanent attachment of myosin to actin is analogous (if not identical) to one of the states in the cycle of interaction which occurs during contraction. X-ray studies of muscles at rest length (where all cross-bridges can interact with actin) and at non-overlap length (where no cross-bridges can interact) show that when a

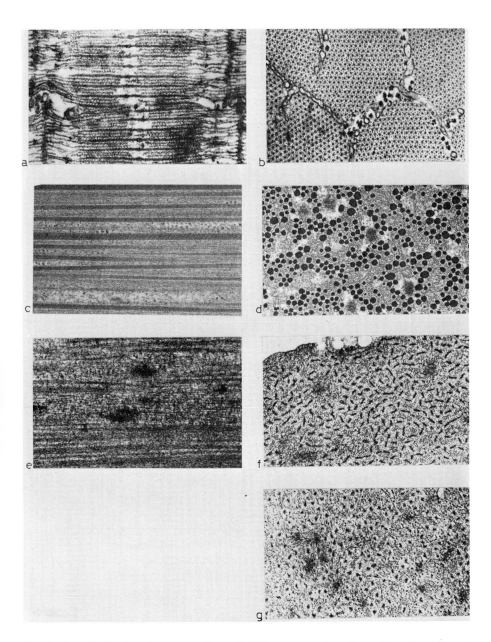

Fig. 1. Longitudinal and cross-section of different muscles. (a and b) Frog sartorius muscle. (c and d) *Anterior byssus* retractor muscle. (e and f) Taenia coli of guinea pig fixed at $0°C$. (g) Cross section of taenia coli of guinea pig fixed at $37°C$. I am indebted to Drs. Huxley, Small and Sobieszek for allowing me to use these micrographs.

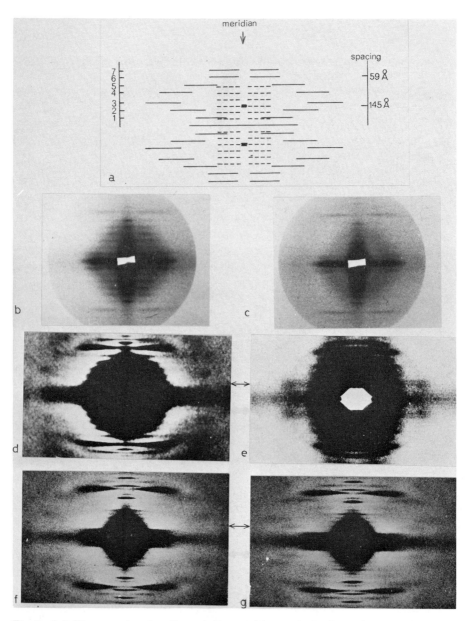

Fig. 2. (a) Diagram showing the relative positions of the layer lines from the actin filaments (thin lines) and the position of the two meridional reflections seen from all muscles (thick spots): the diagram is the same scale as all the X-ray photographs. The myosin layer lines, when they occur, have different spacings from different muscles but they all appear in the region where the dotted lines are shown. The actin layer lines are numbered. (b and c) Diffraction patterns from a frog striated muscle with a sarcomere length of 2.9 μm. (b) at rest, showing the myosin layer lines clearly, (c) contracting: the myosin layer lines are much weaker in (c) than in (b). (d and e) Diffraction patterns from sartorius muscle at rest length showing the actin layer lines. The second layer line (arrowed) is absent in the pattern from the live muscle (d), but clearly visible in the pattern from the rigor muscle (e). (f and g) Diffraction patterns from *Anterior byssus* retractor muscle showing the actin layer lines. (f), relaxed muscle: (g), muscle contracting tonically. The second layer line (arrowed) is stronger during contraction.

muscle passes into rigor, the changes occurring are very similar to those occurring when the muscle contracts. The layer lines disappear completely showing that all the cross bridges have moved away from their positions in the relaxed muscle [8] and the two principal meridional reflections indicate that the myosin filament has increased in length by 1% [6]. (In the patterns from a muscle in rigor at rest-length extra layer lines appear, arising from the structure formed by the cross-bridges attached to the actin filament.)

It seems therefore that the cross-bridge movement detected in the X-ray patterns is not only the motion during the cyclical attachment to and detachment from actin, but is a movement caused by the activation of the muscle. It has been suggested by Haselgrove [6] that the movement of cross-bridges is closely associated with the change in the back-bone of the myosin filament and may be the operation of a myosin-linked control system which operates by holding the cross-bridges back close to the myosin filament in the relaxed muscle and preventing any interaction of myosin with the thin filaments. Such a mechanism would operate in parallel with an actin-controlled system.

Actin Filaments

The structure of the actin-containing thin filaments is helical. G-actin monomers are arranged to form a 2-stranded helical filament which has a pitch of about 730 Å [9]. Tropomyosin molecules lie in the grooves of the actin helix, and a troponin molecule binds to each tropomyosin molecule. Actin-myosin interaction is regulated by Ca^{2+} binding to the troponin molecules which then operate in conjunction with the tropomyosin to permit the interaction of actin and myosin. One troponin-tropomyosin complex will regulate the interaction of seven actin molecules (see for example [10]).

The X-ray pattern given by the actin filaments is much weaker than the myosin pattern discussed above, and the reflections occur much further from the origin (Fig. 2a). Very long X-ray exposures from a relaxed muscle show the series of layer-lines expected from the known actin structure but the 2nd layer-line is absent (Fig. 2d). Computer-modeling studies [7,11,12] suggested that the 2nd layer line is weak because the tropomyosin molecules do not sit exactly at the centre of the groove in the actin helix but sit much closer to one of the chains of monomers than the other. Huxley [7,13] found that when a striated muscle contracts the 2nd actin layer line becomes quite strong and there is no change in the length of the filaments. A similar effect was found in muscles put into rigor at any length [11] (Figs 2d and 2e). This change in the X-ray pattern indicated that the tropomyosin molecules had moved about 10—15 Å nearer the centre of the groove in the actin filaments [7,11,12]. It was then possible to explain how a single tropomyosin-troponin complex could regulate the interaction of seven actin monomers. The mechanism seems to be that in the absence of Ca^{2+} the tropomyosin is held out of the centre of the groove and physically covers the site on every actin monomer to which the S_1 subunit of myosin binds. When the muscle is activated, calcium binds to troponin which then moves the tropomyosin molecule out of the way of the active sites. This explanation is supported by the diffraction patterns from muscles in rigor at non-overlap lengths (where the actin filaments do not have any cross-bridges bound): if Ca^{2+} is present in the muscle, the 2nd layer line is present in the diffraction pattern indicating that the tropomyosin has moved towards the centre of the groove, but in the absence of Ca^{2+}, the 2nd layer line is absent indicating that the tropomyosin molecule has not moved from its position in a relaxed muscle [11].

A further observation which is of importance when studying patterns from smooth muscles (which do not contain troponin) is that if skeletal muscles at rest length are put into rigor in the absence of Ca^{2+} (so that the troponin does not activate the actin filament) the cross-bridges still bind to actin and the tropomyosin moves from the edge towards the centre of the groove [11]. Presumably the cross-bridges are able to bind to the unactivated actin filaments and in so doing push the tropomyosin away from the binding site, but how they do so has not yet been discovered.

"CATCH" MUSCLES

The lamellibranch smooth "catch" muscles (e.g. *Anterior byssus* retractor muscle of *Mytilus*) also work by a sliding filament mechanism, but the structure of the muscle is very different from that of the striated muscles, probably because of their different functions: the catch muscle is primarily intended to maintain high tensions at a fixed length for long periods with a low energy expenditure [14]. Thick filaments are about 30 μm long and have diameters seen in electron microscope cross sections to vary between 200—700 Å in *Anterior byssus* retractor muscle [15] and may be up to 1200 Å in other muscles. They are distributed uniformly throughout the sarcoplasm and are surrounded by many actin filaments each about 11 μm long [15]. In regions in between the thick filaments, the thin filaments pack in a hexagonal lattice with a spacing of about 120 Å and in certain conditions the lattice may have three-dimensional order [16]. It seems unlikely that this packing is necessary for the function of the muscle, it is probably a feature of the way actin filaments tend to pack in vivo.

Thick Filaments

The thick filaments consist of paramyosin, which forms the thick core, and myosin which sits on the surface of the filament with the heads projecting so that they can interact with adjacent actin filaments [17,18]. The paramyosin in the core of the filament may have two functions: (i) to give the filament structural rigidity, and (ii) to modify the myosin ATPase [18] so that the muscle can go into the catch state. (However, paramyosin also occurs in muscles which do not have a catch state.) The paramyosin packs in the filament so that the surface has a well defined surface lattice (the "Bear-Selby-net" [19]), in which successive nodes occur with an axial repeat of 145 Å and the whole structure repeats after 725 Å. Sobieszek [15] and Elliott [20] have recently shown that the myosin packs on the surface of the filament with essentially the same lattice as the paramyosin. Szent-Gyorgyi et al. [18] have shown that the paramyosin core of the thick filaments is a bipolar structure, with opposite ends of the filament polarised in opposite directions. Probably the intact filament with myosin on the surface is bipolar too.

The X-ray pattern from the thick filament is a series of reflections on and close to the meridian, of which the strongest is the meridional reflection at 144 Å. No change can be detected in this pattern when the muscle contracts or when it passes into rigor [21]. It is though that the pattern arises almost entirely from the packing of the paramyosin backbone rather than the myosin surface layer, so the lack of change in the pattern gives us no information about the cross-bridges, but does indicate that there is no significant change in the lattice of the paramyosin structure. Lowy and Vibert [21] have reported that when an *Anterior byssus* retractor muscle is glycerinated to put it into rigor, there are changes in the equatorial pattern that may arise from changes in the lateral packing of the paramyosin molecules in the filament. It is perhaps disappointing that we can find no X-ray evidence for movement of cross-bridges, because biochemical studies by Kendrick-Jones et al. [22] have shown that the actin-myosin interaction is controlled by the myosin molecules and not (as in striated muscle) by the actin filament.

Thin Filaments

The basic structure of the thin filaments is the same as that of striated muscle in that a two-strand helical filament of actin has tropomyosin molecules in each groove, but in catch muscles there is no troponin present [22]. The actin filaments are attached to structures called dense-bodies in such a way that filaments extending from opposite ends of the dense body are polarised in opposite directions [18]. The dense bodies are therefore analogous to the Z lines in striated muscle.

The X-ray diffraction pattern from the thin filaments in relaxed catch muscle contains the same series of layer-lines seen in the pattern from striated muscle (Fig. 2f) [23] and the second layer line is absent from the pattern from resting muscle. During tonic contraction (Fig. 2g) and in rigor, the second layer line is present with about the same intensity as the third layer-line [21,23]. This change from the resting pattern again indicates that the tropomyosin has moved towards the centre of the groove in the actin filament (with no change in the length of the filament) but either the movement is less than the similar movement in skeletal muscle, or a smaller proportion of tropomyosin molecules move. Since catch muscles contain no troponin, the movement of the tropomyosin must have been caused by the heads of the myosin molecules attaching to the thin filaments (as happens when a rest-length skeletal muscle is put into rigor in the absence of Ca^{2+}). Unfortunately smooth muscles cannot be stretched until no actin-myosin overlap exists in order to check this interpretation. Parry and Squire [12] have suggested that the movement of tropomyosin in smooth muscles which contain no troponin may be related to control of a tonic contraction.

VERTEBRATE SMOOTH MUSCLE

There now seems to be general agreement that vertebrate smooth muscles (e.g. taenia-coli of guinea pig) operate on a sliding filament mechanism, since X-ray evidence for longitudinal aggregates of myosin as well as for actin filaments have been obtained from living muscle [24]. Unfortunately, the structure of the thick filaments seen in the electron micrographs depends upon the way the muscle was fixed and equilibrated before fixation (Fig 1f and 1g). Very different results are obtained from muscle fixed at 37° C (e.g. [25]); at room temperature (e.g. [26]) and from muscle fixed near 0° C (e.g. [27]).

Schoenberg has fixed muscles at 37° C in an environment as close as possible to that in which they operate in vivo. She finds both thin and thick filaments, but the thick filaments are irregular in size, shape and in number.

Somlyo, Rice and their co-workers fixed vertebrate smooth muscle with glutaraldehyde at room temperature. They found four principal components in the sarcoplasm which they interpreted as follows: (i) Thick filaments about 160 Å in diameter and 1.5 μm long which have projections from the side. These filaments were thought by Somlyo et al. [26] to be composed only of myosin, and have a structure basically similar to that of the thick filaments in striated muscle. (ii) Thin filaments about 80 Å in diameter which are probably actin filaments and occur around the thick filaments. (iii) Electron dense areas (dense bodies) which Somlyo et al. [26] think are attachment sites of actin filaments (as in catch muscles). (iv) 100-Å thick filaments, which are distinct from the other two filament types and are thought by these workers to be a form of cytoskeleton. They always appear near dense bodies.

Small and co-workers [27,28] cooled muscles to 0° C and equilibrated them in hypertonic solutions to suppress the natural activity of the muscles before fixing them. They found only two principal components which they interpreted as follows: (i) Myosin ribbons. Scattered throughout the sarcoplasm are ribbon shaped structures, about 5 μm long, about 140 Å thick and with widths ranging from 300 to 1100 Å. The ribbons have a backbone structure which consists of the 100 Å filaments described above stacked side by side. Myosin molecules sit on the faces of the ribbons with the heads projecting from the surface: the myosin molecules on opposite faces point in opposite directions so that the ribbon is face-polarised. This arrangement could allow the muscle to operate over a very long range of muscle lengths (see [29] for the length-tension characteristics of taenia coli). The heads of the myosin molecules project from the filament at the nodes of a well defined lattice which is so similar to the surface lattice of myosin in filaments from other muscles that Squire [30] has proposed that the packing is essentially the same in the thick filaments of all muscles. (ii) Thin, actin filaments about 80 Å in diameter and 30 μm long surround each myosin ribbon, and pack into hexagonal arrays about 120 Å apart. 100-Å filaments, dense bodies and thick round filaments are seen much less frequently than in muscles fixed at 37°C, and Lowy, Small and Squire regard these components as being breakdown products of the ribbons, although the ribbons cannot be considered simply as a side-to-side aggregation of the thick filaments because the diameter of the thick filament is too great.

Thus the structure of the myosin containing filaments in vertebrate smooth muscle is still uncertain. The structures seen in muscles fixed at low temperature are much better defined and better ordered than the thick filaments seen in muscles fixed at room temperature or at 37°C, and it seems difficult to explain why the ribbons should have such a precisely ordered structure if they have no structural or functional significance. However, the equilibration conditions in which ribbons are seen are far from the conditions in which the muscle normally operates (indeed a muscle at 0°C will produce only 50% of the tension it produces at 37°C, even with caffeine stimulation) and ribbons are rarely seen in muscles fixed at 37°C.

The X-ray patterns too, indicate that broad ribbons may not be present in contracting muscle. Only two reflections arising from the myosin filaments can be seen in the X-ray diffraction pattern from vertebrate smooth muscle: the meridional reflections at 143 and 72 Å. The shape of the 143-Å reflection indicates that it arises from the ribbon structures rather than the round filaments [31], and the 143-Å reflection is strongest in conditions where ribbons are seen in the microscope. However, the reflection is either very weak or absent in patterns from muscles at 37°C whether the muscle is contracting or is relaxed [25], suggesting that ribbons are not present.

The myosin in vertebrate smooth muscle is obviously very labile [32] and I think it is possible to explain the different appearances of the muscles fixed at different temperatures as follows. The myosin elements have structures like narrow ribbons, but because of the effect of fixation at 37°C they appear as poorly preserved almost round filaments: if the muscle is cooled before fixation then the myosin elements aggregate to form broad ribbons which for some reason are more stable to fixation conditions. Lowy, Poulsen and Pedersen (personal communication) have recently shown that that shape of the 143-Å meridional reflection at 37°C is consistent with a narrow ribbon-like structure which becomes broader as the muscle is cooled.

Because the X-ray reflections are so greatly influenced by the equilibrium conditions of the muscle, it is not possible to draw any conclusions about the movement of the cross-bridges when a muscle contracts.

Thin Filaments

The structure of the actin filament from vertebrate smooth muscle is very similar to the actin filament from the other muscles described above: actin and tropomyosin are wound into a 2-strand helical structure. The layer line pattern given by the actin filaments is the same as that from the other two muscles, with the second layer line being absent from the pattern from relaxed muscle. When the muscle is contracting this layer line is about as strong as the third layer line [23]. So in vertebrate smooth muscle too, contraction is associated with a movement of the tropomyosin in the groove of the thin filament. It is not yet known whether vertebrate smooth muscle contains troponin which may be responsible for this movement, or whether this movement is controlled by the attachment of myosin to actin as in catch muscles.

CONCLUSION

It is now well established that smooth muscles as well as striated muscles operate by a sliding filament mechanism. Thick filaments which have myosin on the surface run parallel to the fibre axis and overlap with thin filaments containing actin.

The thick myosin-containing filaments are markedly different in the different muscle types studied although a common feature of these filaments is that myosin packs with an axial repeat of about 145 Å. Squire ((1971) Nature 233, 457—462) has suggested that the surface lattice of myosin on the surface of the filaments may be basically very similar in all muscles. The X-ray pattern from the myosin structures is very different from muscle to muscle. On one hand, the detailed pattern from relaxed striated muscle changes significantly during contraction and indicates that the whole myosin filament may be responding to activation. On the other hand little information can be gained from the myosin X-ray pattern of contracting smooth muscle other than that the myosin filaments do not change their length.

In contrast to the thick filaments, the actin filament structure seems to be highly conserved from muscle to muscle. The differences between different muscles seem to be confined to the variations in the length of the filament and the presence or absence of troponin. X-ray studies show that during contraction of all types of muscle the tropomyosin molecule moves from a position well away from the centre of the groove to a more central position. In striated muscle this movement is probably caused by the troponin molecules although in smooth muscle it may be caused by the myosin heads binding to the actin filament.

REFERENCES

1. Huxley, H.E. (1969) The mechanism of muscle contraction. Science 164, 1356—1366.
2. Ebashi, S. and Endo, M. (1968) Calcium ion and muscle contraction. Progr. Biophys. Mol. Biol. 18, 123—183.
3. Huxley, H.E. (1963) Electron microscope studies on the structure of natural and synthetic protein filaments from striated muscle. J. Mol. Biol. 7, 281—308.
4. Huxley, H.E. and Brown, W. (1967) The low angle X-ray diagram of vertebrate striated muscle and its behaviour during contraction and rigor. J. Mol. Biol. 30, 383—434.
5. Haselgrove, J.C. and Huxley, H.E. (1973) X-ray evidence for radial cross-bridge movement and for the sliding filament model in actively contracting muscle. J. Mol. Biol. 77, 549—568.
6. Haselgrove, J.C. (1970) X-ray diffraction studies on muscle. Ph.D. Thesis, University of Cambridge.
7. Huxley, H.E. (1972) Structural changes in the Actin and Myosin-containing filaments during contraction. Cold Spring Harbor Symp. Quant. Biol. 37, 361—376.
8. Huxley, H.E. (1967) Recent X-ray diffraction and electron microscope studies of striated muscle. J. Gen. Physiol. 50, 71—83.
9. Hanson, J. and Lowy, J. (1963) The structure of F-actin and of actin filaments isolated from muscle. J. Mol. Biol. 6, 46—60.
10. Bremel, R.D. and Weber, A. (1972) Cooperation within actin filament in vertebrate skeletal muscle. Nature New Biol. 238, 97—101.
11. Haselgrove, J.C. (1972) X-ray evidence for a conformational change in the actin-containing filaments of vertebrate striated muscle. Cold Spring Harbor Symp. Quant. Biol. 37, 341—352.
12. Parry, D.A. and Squire, J.M. (1973) Structural role of tropomyosin in muscle regulation: Analysis of the X-ray diffraction patterns from relaxed and contracting muscles. J. Mol. Biol. 75, 33—35.
13. Huxley, H.E. (1971) Structural changes during muscle contraction. Biochem. J. 125, 85P.
14. Lowy, J. and Millman, B.M. (1963) The contractile mechanism of the *Anterior byssus* retractor muscle of *Mytilus edulis*. Phil. Trans. Roy. Soc. 246, 105—148.
15. Sobieszek, A. (1973) The fine structure of the contractile apparatus of the Anterior Byssus retractor muscle of Mytilus edulis. J. Ultrastruct. Res. 43, 313—343.
16. Lowy, J. and Vibert, P.J. (1967) Structure and organisation of actin in a molluscan smooth muscle. Nature 215, 1254—1255.
17. Hanson, J. and Lowy, J. (1964) The structure of molluscan tonic muscles. In Biochemistry of muscle contraction. Gergely, J., (ed.) pp. 400—409, Churchill London.
18. Szent-Gyorgyi, A.G., Cohen, C. and Kendrick-Jones, J. (1971) Paramyosin and the filamenti of molluscan "catch muscles". J. Mol. Biol. 56, 239—258.

138

19. Bear, S. and Selby, C.C. (1956) The structure of paramyosin fibrils according to X-ray diffraction. J. Biophys. Biochem. Cytol. 2, 55—69.
20. Elliott, A. (1974) The arrangement of myosin on the surface of paramyosin filaments in the white adductor muscle of *Crassostrea Angulata*. Proc. Roy. Soc. Lond. Ser. B 186, 53—66.
21. Lowy, J. and Vibert, P.J. (1972) Studies of the low-angle X-ray pattern of a molluscan smooth muscle during tonic contraction and rigor. Cold Spring Harbor Symp. Quant. Biol. 37, 353—359.
22. Kendrick-Jones, J., Lehman, W. and Szent-Györgyi, A.G. (1970) Regulation in molluscan muscles. J. Mol. Biol. 54, 313—326.
23. Vibert, P.J., Haselgrove, J.C., Lowy, J. and Poulsen, F.R. (1972) Structural changes in actin-containing filament of muscle. J. Mol. Biol. 71, 757—767.
24. Lowy, J., Poulsen, F.R. and Vibert, P.J. (1970) Myosin filaments in vertebrate smooth muscle. Nature 225, 1053—1054.
25. Shoenberg, C.F. and Haselgrove, J.C. (1974) Filaments and ribbons in vertebrate smooth muscle. Nature 249, 152—154.
26. Somlyo, A.P., Devine, C.E., Somlyo, A.V. and Rice, R.C. (1973) Filament organisation in vertebrate smooth muscle. Phil. Trans. Roy. Soc. Lond. Ser. B 265, 223—229.
27. Small, J.V. and Squire, J.M. (1972) Structural basis of contraction in vertebrate smooth muscle. J. Mol. Biol. 67, 117—149.
28. Lowy, J. and Small, V.J. (1970) The organisation of myosin and actin in vertebrate smooth muscle. Nature 227, 46—57.
29. Lowy, J. and Mulvany, M.J. (1973) Mechanical properties of guinea-pig taenia coli muscles. Acta Physiol. Scand. 88, 123—136.
30. Squire, J.M. (1971) General model for the structure of all myosin containing filaments. Nature 233, 457—462.
31. Lowy, J., Vilbert, P.J., Haselgrove, J.C. and Poulson, F.R. (1973) The structure of the myosin elements in vertebrate smooth muscles. Phil. Trans. Roy. Soc. Lond. Ser. B 265, 191—196.
32. Shoenberg, C.F. (1973) The influence of temperature on the thick filaments of vertebrate smooth muscle. Phil. Trans. Roy. Soc. Lond. Ser. B 265, 197—202.

Comparative Physiology — Functional Aspects of Structural Materials
Eds L. Bolis, H.P. Maddrell and K. Schmidt-Nielsen
© North-Holland Publishing Company — 1975 — Amsterdam

Insect fibrillar muscle and the problem of contractility

J.W.S. PRINGLE

A.R.C. Unit of Muscle Mechanisms and Insect Physiology, Department of Zoology, South Parks Road, Oxford (U.K.)

INTRODUCTION

It is now 25 years since it was first reported that rhythmic mechanical activity of the striated flight muscles of certain higher insects was not produced by the arrival of synchronous volleys of impulses in their motor nerves [1,2]. Many of the insect species which show the phenomenon are small, but with the discovery of a few, mainly tropical species from which preparations could be made of the isolated muscles, the peculiar physiological features of insect fibrillar muscle, which are responsible for oscillation in the intact animal, have been reduced to two:

(1) Great resistance to stretch in the relaxed condition, stresses of up to $0.5 \text{ kg} \cdot \text{cm}^{-2}$ being generated by a strain of about 5%, and

(2) a direct influence of length on the tension generated by active muscle, with a delay between length changes and the resulting active tension (stretch activation).

The early work is reviewed by Pringle [3].

The most useful preparation is the dorsal longitudinal muscle of giant water bugs of the family Belostomatidae. The largest species, *Lethocerus maximus*, which occurs in Trinidad and South America, is 30 cm long; the fibres of its dorsal longitudinal muscle have a length of 2 cm and extend from one end of the muscle to the other with a uniform diameter of 70 μm. Using this muscle from a number of related species, it has been shown, first, that both these properties reside in the contractile myofibrils and not in any peculiarity of the excitation of the plasma membrane of the fibre or in the nature of the excitation-contraction coupling process. Fibres in which the lipoprotein membrane systems of the cell have been dispersed by glycerol extraction and/or treatment with detergents still behave in a quantitatively similar way to the living fibre [4]. Secondly, it is now known that in respect

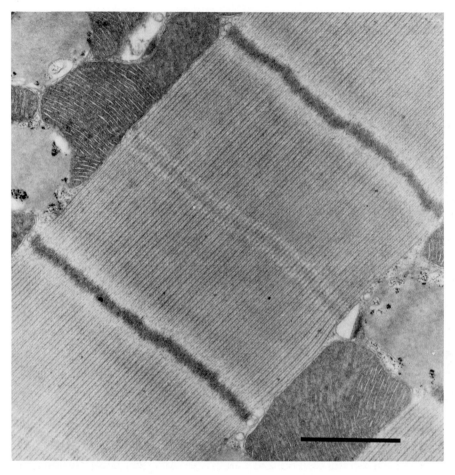

Fig. 1. Longitudinal section of dorsal longitudinal flight muscle of *Lethocerus*; scale 1 μm. (Electronmicrograph by Barbara Luke.)

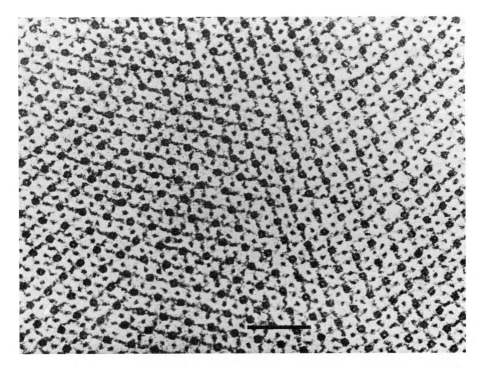

Fig. 2. Transverse section of part of a myofibril of *Lethocerus* in rigor; scale 100 nm (similar to Reedy [36]). This work showed that the myosin molecules are arranged in the filament in a left-handed helix and the actins in a right-handed helix.

of both their peculiar properties, insect fibrillar muscles differ only quantitatively from other types of muscle. Relaxed fibres from various types of non-fibrillar insect muscle show all gradations of resistance to stretch from the extreme plasticity of frog and rabbit muscles to the strong visco-elasticity of insect fibrillar muscles [5,6]. Stretch activation has been found to be present in a non-fibrillar insect muscle [7] and in rabbit psoas, frog semitendinosus and even rabbit heart muscle [8], though it is not developed to the extent that would enable these muscles to show oscillatory mechanical activity in their natural situation in the body.

In their general fine-structural organization, the myofibrils of insect fibrillar muscle are similar to those of vertebrate striated muscle. They are, however, thicker (diameter up to 3 μm), have a very short I-band with almost complete overlap of actin and myosin filaments and are extremely regular; every sarcomere along a fibre is identical with its neighbours and within a sarcomere the filaments are arranged both transversely and longitudinally with almost crystalline regularity (Figs 1 and 2). For a recent review of fine structure, see Pringle [9].

THE MOLECULAR MECHANISM OF CONTRACTION

This paper deals mainly, not with the reasons why resistance to stretch and stretch activation are better developed in insect fibrillar muscles, but with the use which has been made of this tissue for the better understanding of the problems of contractility at the molecular level. There are good reasons for thinking that the transduction of chemical to mechanical energy is performed in an essentially similar manner in all muscles and also in the many non-muscular contractile systems that involve the interaction of filamentous actin with myosin-like molecules [10]. Knowledge of the nature of this mechanism has progressed by making use of the opportunities presented by a variety of tissues.

It was the extreme regularity of insect fibrillar muscle that provided the first opportunity. Using a combination of electronmicroscopy and X-ray diffraction techniques on glycerol-extracted fibres in a controlled chemical environment, Reedy et al. [11] showed that in relaxing solution (MgATP present but $Ca^{2+} < 10^{-9}$ M), the cross-bridges, containing the heads of the myosin molecules, are oriented primarily at right-angles to the fibre axis with a periodicity determined by their origins on the myosin filament; by contrast, in rigor solution (MgATP absent), the bridges are attached to the actin filaments at an angle and accentuate the actin periodicity (Fig. 3). This was the first good demonstration of the presence of links between the filaments in mechanical rigor; the configurational change occurring in the bridges has

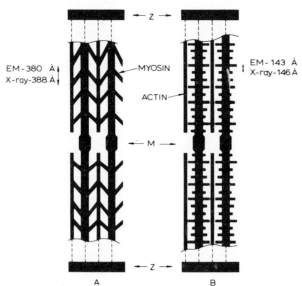

Fig. 3. Diagrams of single layers of actin and myosin filaments showing cross-bridge positions found to be characteristic of (a) rigor and (b) relaxed state of *Lethocerus* myofibrils (from Reedy et al. [11]).

been taken to represent states which the bridges assume at certain instants during the cycle of movement which they perform during activity.

The basic feature of contractility is the active process of interaction between myosin and actin which has come to be called the cross-bridge cycle. The mechanical force causing the filaments to slide relative to each other is generated by a cycle of events involving attachment of the head of the myosin molecule to the actin filament, a configurational change in which energy is released, detachment and recovery. At some point during this cycle, MgATP is hydrolysed. This dynamic interaction between myosin and actin takes place at maximal rate when the pure proteins are mixed in solution, provided that MgATP is present and the actin is in the filamentous form which it assumes in the presence of salt. The myosin can react either as a monomer at high salt concentration or as a filament at the physiological salt concentration of about 0.1 M. In solution, no mechanical forces can be generated and the work is degraded to heat.

Quite distinct from the cross-bridge cycle are the processes in intact myofibrils whereby the active mechanochemical events are regulated. These involve other thin filament proteins, tropomyosin and the troponins, and, in some muscles, a direct effect of Ca^{2+} on certain parts of the myosin molecule. The effect of these additional features is that mechanochemical activity is controlled in the otherwise constant chemical environment of the muscle cell by micromolar concentrations of Ca^{2+} released from the sarcoplasmic reticulum at the last stage of the excitation-contraction coupling process and, in insect fibrillar muscle, also to a significant extent by a direct effect of length changes.

EVIDENCE ABOUT REGULATION MECHANISMS

In vertebrate striated muscle the only effective control of actomyosin interaction is that mediated by Ca^{2+} and the only way in which Ca^{2+} is known to affect the proteins of the myofibril is through the complex located in the thin filaments. Ca^{2+} binds at micromolar concentration to troponin C. This releases the inhibition of actomyosin ATPase exercised by troponin I. In the presence of the third component, troponin T, the tropomyosin is caused to change its position relative to the double helix of actin molecules in the filament, exposing the myosin-binding site on actin and allowed that attachment phase of the cross-bridge cycle to occur. The evidence for these events is reviewed by Perry [12] and Huxley [13].

The muscles of molluscs do not contain the components of the troponin complex, but their activity is nevertheless regulated by Ca^{2+}. The mechanism involves a direct effect of Ca^{2+} on myosin and is lost when one of the two light chains of the molecule is removed [14]. In many groups of animals including *Annelids*, *Crustacea* and *Echinoderms* [15] and also in insects

[16], both regulatory mechanisms are present; they are complementary to each other and combine to produce the large difference in ATPase activity between rest and activity.

As indicated above, a beginning has been made in understanding how Ca^{2+} control operates through a change in the patterns of association of the proteins of the thin filament. X-ray diffraction has also provided evidence of changes occurring in the thick filaments. The distribution of intensities along the equator of the diffraction pattern provides evidence of the distribution of density in the cross-section of the myofibrils. In this way the cross-bridges in relaxed rabbit muscle have been shown to extend out only a short distance from the myosin filament and to be about 70 Å away from the actin filament; by contrast, in rigor, they are clearly attached to the actin [17,18]. In active muscle, stimulated at its normal sarcomere length where there is considerable overlap between actin and myosin filaments, the distribution of intensities along the equator shows that in rabbit muscle about half the bridges are in the vicinity of the actin filaments, though not necessarily all attached to them. In muscles stimulated at sarcomere lengths at which there should be no overlap, some movement of the cross bridges can still be detected both by the equatorial pattern and from the diminution of intensity of the layer line generated by their axial order. Huxley [13] suggests, however, that this might be explained by cooperative effects between bridges rather than by a direct recognition by the myosin filaments that the muscle is active, a phenomenon which would be inconsistent with the present evidence that there is no calcium control of myosin in vertebrate muscle.

Insect fibrillar muscle differs from vertebrate muscle in that there is a second mechanism of regulation, stretch activation superimposed on the normal Ca^{2+} control. In fibrillar myofibril suspensions, increase in Ca^{2+} concentration produces only a small increase in ATPase activity [19]; in unstretched fibres, there is also only slight active tension. Full activity demands also that the fibres be stretched. This makes it possible to separate the direct effects of Ca^{2+} from those associated with the operation of the cross-bridge cycle.

Fig. 4 shows X-ray diffraction pictures obtained from insect fibrillar muscle in relaxation and rigor. Measurements of Spot M show that when Ca^{2+} is added to relaxed, unstretched fibrillar muscle, no axial movement of the cross-bridges takes place. On the other hand, the equatorial distribution of intensities suggests that 10—20% of the bridges move out to the actin filaments (review by Miller and Tregear [20]); they do not, however, attach to the actin as in rigor, since this would lead to a decrease in the meridional intensity which is not observed. This suggests a direct influence of raised Ca^{2+} concentration on the myosin filament, since long-range forces would be required to produce the effect through a change in the organization of the thin filaments.

During stretch activation (a more complete explanation of these experi-

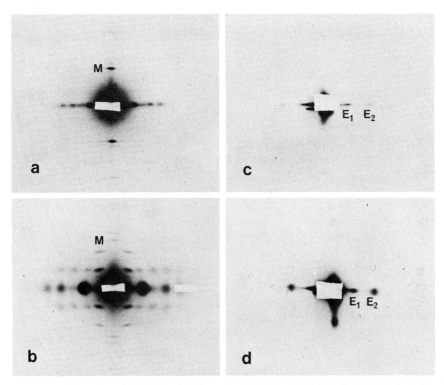

Fig. 4. X-ray diffraction of insect fibrillar muscle in relaxation (a, c) and rigor (b, d). The scale and exposure in (a) and (b) were different from those in (c) and (d). Axial movement of the cross-bridges is indicated by reduction in intensity of Spot M (meridian), which is due to the regular spacing of the heads of molecules along the myosin filament. An increase in the ratio of intensities of E_2/E_1 (equator) indicates radial movement of bridges towards the actin filaments. (Unpublished photographs by Miller, A. and Tregear, R.T.)

ments is given in the next section) there is no change in the distribution of intensities along the equator of the diffraction pattern, showing that, on average, no more bridges move out to the vicinity of the actin filaments during activity. Stretch activation therefore operates either through the thin filaments or through some different influence on the ability of the myosin heads to attach to actin.

The mechanism of stretch activation is not understood, but it is convenient to mention here the evidence that it does not work through a change in the binding constant for Ca^{2+} of any component of the myofibril. Marston and Tregear [21] showed that there was no significant change in Ca^{2+} binding in fibrillar muscle fibres when these were stretched by 5%, although there was a 30% increase in ATPase activity. An earlier report of a positive effect [22] was probably due to defective technique. The three ways of activating ATPase activity, Ca^{2+} binding to myosin, Ca^{2+} binding to thin filaments and stretch evidently all act in a synergistic manner.

EVIDENCE ABOUT THE NATURE OF THE CROSS-BRIDGE CYCLE

Movement of Cross-Bridges

The slowness of the process of chemical fixation needed before fine structure can be observed with the electron microscope eliminates this technique as a method of finding out what happens during the dynamic process involved in contractile activity. X-ray diffraction does not suffer from this limitation and most of the evidence about dynamic aspects of the cross-bridge cycle have been obtained with this technique. Using frog muscle, Huxley and Brown [23] and Haselgrove [24] were able to conclude from measurements of intensity changes in various diffraction spots that during active contraction the cross-bridges are in movement axially and either azimuthally or radially in relation to the filament axis. The fact that diffuse scatter increases in actively contracting frog muscle relative to relaxed muscle suggested that the movement of the various cross bridges was unsynchronized. This is what would be expected in a muscle whose function is to generate steady tensions or a smooth sliding movement.

A similar teleological argument applied to insect flight muscles would suggest that in the oscillatory contraction which occurs in life there should be some synchronization of the cross-bridge cycles. For the optimum performance of oscillatory mechanical work, energy has to be fed into the cycle only during the shortening phase and the mechanochemical conversion process will be most efficient when the frequency of mechanical oscillation is the same as the turn-over frequency of the actomyosin enzyme system. Biochemical evidence that this is the case was reviewed by Pringle [25].

If partial synchronization of the cross-bridge cycles does occur, then there should be a difference in the number of attached bridges at different phases of the oscillation. Because of differences between the axial repeats along the myosin and actin filaments, attached cross-bridges cannot retain the regular spacing responsible for the intense meridional spot present in the diffraction picture of relaxed muscle (Fig. 4); the intensity of this spot will be further reduced if the bridges which are attached during activity are angled in the same way as is seen during attachment in rigor. Tregear and Miller [26] showed that the intensity of the meridional spot is indeed reduced at the phase of the oscillation cycle at which tension is maximal (Fig. 5). On the assumption that the bridges which attach to actin are the same bridges that were shown to move out from the myosin filament on Ca^{2+} activation, they concluded that about 15% of the total were simultaneously attached during active oscillation and that the axial displacement of these bridges is about 100 Å [20].

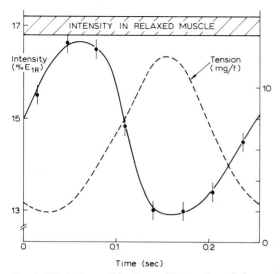

Fig. 5. Variation with time of intensity of the main meridional spot and tension during oscillation of activated *Lethocerus* muscle fibres (from Miller and Tregear [20]).

These experiments were done using amplitudes of oscillation at which a large amount of mechanical work is performed. The fact that only 15% of the bridges are simultaneously attached does not mean that only this small proportion is contributing energy to the oscillation. An oscillation amplitude of 3% corresponds to a sliding movement of 300 Å in each half-sarcomere, which is too large to be produced by the axial displacement of a single cross-bridge. There is also independent evidence that the duration of the actin-myosin interaction is considerably less than that of the half-cycle of oscillation. Provided that the turn-over frequency of the cross-bridge cycle is the same as that of the oscillation, bridges which are attached during any part of the shortening phase will contribute usefully to the work output. Exact simultaneity between the cycles of operation is thus not a requirement for mechanochemical efficiency during oscillation.

Differential Effects on Different Points in the Cross-Bridge Cycle

In order to interpret experimental results in terms of effects occurring at different points in the cross-bridge cycle, it is necessary to have a model of its operation. The first attempt to define such a model was that of Huxley [27] and most subsequent models have been of the same type [28]. The number of attached bridges (n) may vary both with time (t) and with the amount of sliding which has occurred (x). A differential equation for the making and breaking of attachments can therefore be written:

$$\frac{\partial n(x,\ t)}{\partial t} = f(x,\ t)\ \{1 - n(x,\ t)\} - g(x,\ t)\ n(x,\ t) \tag{1}$$

where $f(x,t)$ and $g(x,t)$ are, respectively, the probabilities that a detached bridge will attach and that an attached bridge will detach.

In order to use mathematical models of this sort to interpret mechanical results, it is necessary to make some assumptions about the generation of tension. The simplest assumption is that all attached bridges generate a quantum of tension; if, then, only isometric conditions are considered, the formula reduces to

$$\frac{dn}{dt} = f(1 - n) - gn \tag{2}$$

Since stretch activation of insect fibrillar muscle can be measured accurately for extremely small changes of length (0.05% peak-to-peak), Abbott [29] used this formulation to analyse the results of experiments in which the physical and chemical conditions were varied. If stretch is assumed to activate by its influence on the attachment probability, f, delayed tension generation follows first-order kinetics, as Eqn 2 demands, and the rate constant of the delay measures $f + g$. This rate constant
(1) has a high temperature coefficient,
(2) is increased as Ca^{2+} is raised,
(3) is unaffected by the mean stretch of the fibres,
(4) is decreased by ADP.

Combined with other evidence (discussed above) that even with large amounts of activation the proportion of attached bridges is small ($f \ll g$), this appears to mean that these influences are affecting g, the probability of detachment of attached bridges. No such direct method of measuring this parameter in the cross-bridge cycle is available from experiments with other types of muscle.

At larger amplitudes of length change, non-linear effects become significant and modelling becomes much more difficult; there is no general agreement about the best formulation. White and Thorson [28] give the most complete review of the various possibilities.

Another method of measuring the rate constant of the detachment process in insect fibrillar muscle was developed by Pybus and Tregear [30]. Using glycerol-extracted fibres at various degrees of extension, with or without simultaneous low amplitude length modulation, they determined the relationship between tension and the rate of production of ADP; since this was linear, they could obtain a measure of the "tension cost" in terms of ATP hydrolysed. They also determined the optimum frequency for the performance of oscillatory work in the same preparations. The two measurements varied similarly with temperature, and were therefore likely to be measures of the same parameter. The tension cost was always greater when the muscle was oscillated, suggesting that during the cross-bridge cycle, mechanical energy is liberated rapidly on attachment and stored locally in a spring which shortens and transfers its energy to the sliding of the filaments when shortening is allowed to occur. Since the tension in a spring is proportional to its extension, oscillating the fibres so that it is allowed to shorten will reduce its average tension during its lifetime and so decrease the total tension; the tension cost is therefore increased. The data were also used to make an independent estimate of the proportion of bridges simultaneously attached; the value of 10—20% obtained in this way is similar to that obtained from the X-ray experiments already described.

CORRELATION OF CHEMISTRY AND MECHANICS

The ultimate objective in the study of the molecular mechanism of contraction is to correlate the chemistry of MgATP hydrolysis with the generation of tension and the sliding of the filaments. Important recent developments in this field have come from "stopped-flow" and "quench-flow" experiments using pure myosin from rabbit muscle; by these methods the kinetics of the enzyme reaction have been nearly completely described by Lymn and Taylor [31] and by Trentham and his colleagues [32,33]. During the formation of complexes between the myosin enzyme and its substrate ATP, fluorescence is enhanced and by careful planning of the experimental conditions it is possible to formulate seven steps in the hydrolysis reaction and to measure the equilibrium constants and forward and backward rate constants for each transition. Some of the results are summarized in the inner circle of Fig. 6.

The steps in the reaction are numbered 1—7. Under the conditions at which measurements were made (note that the pH is not the physiological pH of the muscle cell), the slowest step is Reaction 4, which has a forward rate constant of 0.06 s^{-1}.

It is not at present possible to make more than a beginning with the utilization of this information to interpret the changes occurring in intact myofibrils. Myosin ATPase is activated by actin in vitro and this no doubt reflects the interaction that takes place in myofibrils from the attachment-detachment cycle indicated by mechanical results. In order to identify possible experimental attack on the problem, an outer circle has been included in Fig. 6, representing the theoretically possible steps in the reaction when actin and myosin are associated; no values can be assigned to equilibrium or rate constants on this outer circle.

The following experimental observations can be related to this diagram:

(1) In the absence of MgATP, actin and myosin form a tight complex. This corresponds to rigor in the intact myofibril.

(2) When MgATP is added to a glycerol-extracted fibre in rigor in the absence of Ca^{2+}, the fibre relaxes and behaves as if the cross-bridges were detached [34]. This is shown in Fig. 6 as a thicker arrow connecting the two

150

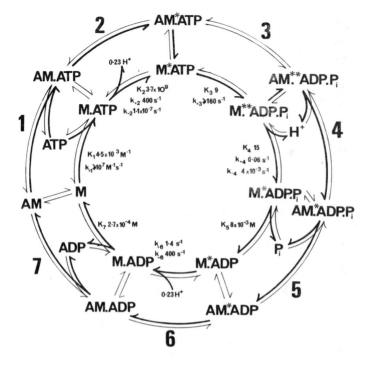

MYOSIN AND ACTOMYOSIN ATPase
0·1M KCl pH 8·0

Fig. 6. Inner circle: kinetics of myosin ATPase, showing equilibrium constants (K) and forward and backward rate constants (k_+, K_-); asterisks indicate enhanced fluorescence. Outer circle: theoretically possible steps in the reaction when myosin and actin are combined. In the inner circle, thickened arrows show the forward reaction; in the outer circle and radial connexions, thickened arrows show the probable course of the reaction in intact myofibrils (data from Trentham et al. [33]; Bagshaw and Trentham [32]; Trentham, D.R., unpublished).

circles between Steps 2 and 3, on the assumption that the "early burst" of H^+ release seen with myosin is also present in actomyosin.

(3) In relaxed muscle, the bound nucleotide is very largely ADP, even when the nucleotide applied to the fibres is ATP [35]. The attachment step is therefore shown as a thickened radial arrow between Steps 3 and 4, but this is a hypothesis. It is possible that, of the three products of ATP hydrolysis, the H^+ and even the inorganic phosphate are released before attachment takes place; in that case, the radial arrow should be between Steps 4 and 5 or between Steps 5 and 6. But the activation of enzymic activity by actin suggests that the location of attachment between Steps 3 and 4, immediately before the rate-limiting step in myosin kinetics, is correct. As stated above, there is evidence that ADP reduces the probability of detachment, so that Step 7 must occur with actin and myosin associated.

This diagrammatic representation of the chemistry of the reaction helps to pinpoint the need for methods of correlating chemical and mechanical events in intact myofibrils with a time resolution great enough to enable the individual steps in the reaction to be identified. At the present time, the only methods that hold out hope of success are the use of ATP analogues that halt the cycle at a particular point, the use of fluorescent probes that can be used in the intact system and possibly measurements of electrical conductivity to detect the release of hydrolysis products. A full discussion of current attempts to use these methods is beyond the scope of this article.

REFERENCES

1. Pringle, J.W.S. (1949) The excitation and contraction of the flight muscles of insects. J. Physiol. London 108, 226—232.
2. Roeder, K.D. (1951) Movement of the thorax and potential changes in the thoracic muscles of insects during flight. Biol. Bull. Woods Hole, 100, 95—106.
3. Pringle, J.W.S. (1967) The contractile mechanism of insect fibrillar muscle. Prog. Biophys. Mol. Biol. 17, 1—60.
4. Abbott, R.H. and Chaplain, R.A. (1966) Preparation and properties of the contractile elements of insect fibrillar muscle. J. Cell. Sci. 1, 311—330.
5. Buchthal, F. and Weis-Fogh, T. (1956) Contribution of the sarcolemma to the force exerted by resting muscle in insects. Acta Physiol. Scand. 35, 345—364.
6. Machin, K.E. and Pringle, J.W.S. (1959) The physiology of insect fibrillar muscle: II. The mechanical properties of a beetle flight muscle. Proc. Roy. Soc. London Ser. B 151, 204—225.
7. Aidley, D.J. and White, D.C.S. (1969) Mechanical properties of glycerinated fibres from the tymbal muscles of a Brazilian cicada. J. Physiol. London 205, 179—192.
8. Rüegg, J.C., Steiger, G.J. and Schädler, M. (1970) The mechanical activation of the contractile system of skeletal muscle. Pflügers Arch. 319, 139—145.
9. Pringle, J.W.S. (1972) Arthropod muscle. In The Structure and Function of Muscle (Bourne, G.H., ed.) pp. 491—541, Academic Press, New York.
10. Pollard, T.D. and Weihing, R.R. (1974) Actin and myosin and cell movement. From CRC Critical Reviews of Biochemistry (Pollard, T.D. and Weihing, R.R., ed.), Vol. 2, pp. 1—65, CRC Press, Cleveland, Ohio.
11. Reedy, M.K., Holmes, K.C. and Tregear, R.T. (1965) Induced changes in orientation of the cross-bridge of glycerinated insect flight muscle. Nature 207, 1276—1280.
12. Perry, S.V. (1973) The control of muscular contraction. Symp. Soc. Exp. Biol. 27, 531—550.
13. Huxley, H.E. (1973) Structural changes in the actin- and myosin-containing filaments during contraction. Cold Spring Harbor Symp. Quant. Biol. 37, 361—376.
14. Szent-Györgyi, A.G., Szentkiralyi, E.M. and Kendrick-Jones, J. (1973) The light chains of scallopmyosin as regulatory subunits. J. Mol. Biol. 74, 179—203.
15. Lehman, W., Kendrick-Jones, J. and Szent-Györgyi, A.G. (1973) Myosin-linked regulatory systems: comparative studies. Cold Spring Harbor Symp. Quant. Biol. 37, 319—330.
16. Lehman, W., Bullard, B. and Hammond, K. (1974) Calcium-dependent myosin from insect muscles. J. Gen. Physiol. 63, 553—563.
17. Huxley, H.E. (1968) Structural difference between resting and rigor muscle; evidence from intensity change in the low-angle equatorial X-ray diagram. J. Mol. Biol. 37, 507—520.

18. Haselgrove, J. and Huxley, H.E. (1973) X-ray evidence for radial cross-bridge movement and for the sliding filament model in actively contracting skeletal muscle. J. Mol. Biol. 77, 549—568.
19. Maruyama, K., Pringle, J.W.S. and Tregear, R.T. (1968) The calcium sensitivity of ATPase activity of myofibrils and actomyosins from insect flight and leg muscles. Proc. Roy. Soc. London Ser. B. 169, 229—240.
20. Miller, A. and Tregear, R.T. (1971) X-ray studies on the structure and function of vertebrate and invertebrate muscle. In Contractility of muscle cells and related processes (Podolsky, R.J., ed.), pp. 205—228, Prentice-Hall, Eaglewood Cliffs.
21. Marston, S. and Tregear, R.T. (1974) Calcium binding and the activation of fibrillar insect flight muscle. Biochim. Biophys. Acta 347, 311—318.
22. Chaplain, R.A. (1966) The effect of Ca^{2+} and fibre elongation on the activation of the contractile mechanism of insect fibrillar flight muscle. Biochim. Biophys. Acta 131, 385—392.
23. Huxley, H.E. and Brown, W. (1967) The low angle X-ray diagram of vertebrate striated muscle and its behaviour during contraction and rigor. J. Mol. Biol. 30, 383—434.
24. Haselgrove, J. (1970) X-ray diffraction studies on muscle. Ph.D. Thesis, Cambridge University.
25. Pringle, J.W.S. (1968) Mechano-chemical transformation in striated muscle. Symp. Soc. Exp. Biol. 22, 67—86.
26. Tregear, R.T. and Miller, A. (1969) Evidence of cross-bridge movement during contraction of insect flight muscle. Nature 222, 1184—1185.
27. Huxley, A.F. (1957) Muscle structure and theories of contraction. Progr. Biophys. 7, 255—318.
28. White, D.C.S. and Thorson, J. (1973) The kinetics of muscle contraction. Prog. Biophys. Mol. Biol. 27, 175—255.
29. Abbott, R.H. (1973) An interpretation of the effects of fiber length and calcium on the mechanical properties of insect flight muscle. Cold Spring Harbor Symp. Quant. Biol. 37, 647—654.
30. Pybus, J. and Tregear, R.T. (1973) Estimates of force and time of actomyosin interaction in an active muscle and of the number interacting at any one time. Cold Spring Harbor Symp. Quant. Biol. 37, 655—660.
31. Lymn, R.W. and Taylor, E.W. (1970) Transient state phosphate production in the hydrolysis of nucleoside triphosphates by myosin. Biochemistry 9, 2975—2983.
32. Bagshaw, C.R. and Trentham, D.R. (1973) The reversibility of adenosine triphosphate cleavage by myosin. Biochem. J. 133, 323—328.
33. Trentham, D.R., Bardsley, R.G., Eccleston, J.F. and Weeds, A.G. (1972) Elementary processes of the magnesium ion-dependent ATPase activity of heavy meromyosin. Biochem. J. 126, 635—644.
34. White, D.C.S. (1970) Rigor contraction and the effect of various phosphate compounds on glycerinated insect flight and vertebrate muscle. J. Physiol. London 208, 583—605.
35. Marston, S. (1973) The nucleotide complexes of myosin in glycerol-extracted muscle fibres. Biochim. Biophys. Acta 305, 397—412.
36. Reedy, M.K. (1968) Ultrastructure of insect flight muscle. I. Screw sense and structural grouping in the rigor cross-bridge lattice. J. Mol. Biol. 31, 155—176.

PART 3

ENERGY DEMANDS ON ANIMALS

Comparative Physiology — Functional Aspects of Structural Materials
Eds L. Bolis, H.P. Maddrell and K. Schmidt-Nielsen
© North-Holland Publishing Company — 1975 — Amsterdam

The energy cost of active transport

R.D. KEYNES

Physiological Laboratory, University of Cambridge, Cambridge (U.K.)

We have an energy crisis; and in this crisis, possibly the single most important element is the energy cost of transport. In preparing this paper I have amused myself by trying to attach a few numbers to this statement, and I hope that my provisional conclusions will be of some interest, even if I begin by trespassing into an area somewhat outside the normal boundaries of comparative physiology.

First, then, I want to present a few facts about the energy cost of the transport mechanism that has brought us to this delightful place, considered in relation to our own metabolism. The standard sedentary Western European man, who weighs 70 kg and has a body surface area of 1.73 m^2, can be taken to have a basal metabolism of 1800 kcal/day, and a total metabolism (as long as his physical activity is limited to listening to lectures) of about 2600 kcal/day. His personal energy expenditure on food, ignoring for the moment the large additional energy cost of producing that food, is therefore some 1100 kWh in a single year. Now consider his personal motor car. An ordinary European car represents around 25 000 kWh of energy expenditure just on the materials used for its construction, while the total energy cost of running it for 12 000 miles amounts to a further 25 000 kWh. My average man may therefore dissipate over 25 times as much energy in private transport as he does in private metabolism. There are, of course, other large items in his energy budget, heating his house being another important one, at least in Northern Europe; but even if he depends mainly on public facilities for his daily movements between home and office, transport remains the predominant item unless he propels himself entirely by his own muscles. As far as our society is concerned, how to cut down on the energy cost of transport is a matter for economists, sociologists and politicians rather than comparative physiologists. All we can do is to point out the striking savings that can be achieved by walking or bicycling to work!

Now let me get down to some physiology, and pose the question: How much of our basal metabolism is devoted to active transport? This cannot be answered without first partitioning our energy expenditure between the various basic functional activities, and to conduct such an operation with any pretence of accuracy is surprisingly difficult. I naively supposed when I set out to do it that the total energy expenditure of the different internal organs of my average man would often have been measured, so that there would be plenty of relevant information in the literature, but this does not seem to be the case. One has in fact to draw on a few scattered observations made under far from standardised conditions for a variety of animals, and the values assembled in the table below must be regarded as approximate in the extreme. My principal source of information has been the table published in the 7th edition (1972) of Geigy's "Tables Scientifiques", taken from Brožek and Grande [1], and itself based on a small number of experimental determinations of the metabolism of individual organs by indirect calorimetry. It seemed advisable, however, to modify some of their figures in the light of more recent observations depending on direct as well as indirect calorimetry. Thus the percentage of basal metabolism attributed to the liver ranges from 26% in man [1] to 13% in a dog [2], and 15 and 11% respectively in pigs and sheep (Webster, A.J., personal communication; and see [3]). I have chosen 18% as an admittedly arbitrary compromise. My figure for the metabolism of the remainder of the digestive organs is similarly a compromise between the estimate of Durotoye and Grayson [2] of 20% for a dog, and the unpublished determinations of Webster of 22% for a fasting sheep, and 30 and 36% for pigs and sheep on a twice maintenance ration. That the percentages given in the table should add up almost exactly to 100 is definitely fortuitous.

We next need to estimate the fraction of the energy consumption of each organ that can plausibly be ascribed to active transport. Ideally, we should calculate the precise amount of osmotic work performed in each case, but this is a counsel of perfection, applicable to very few tissues. As an example of this approach, the human erythrocyte can be taken to maintain an active transport of 2 mmol K^+ inwards against a concentration gradient of 27.4, and an active transport of 3 mmol Na^+ outwards against a concentration gradient of 6.3, both quantities being expressed per l cells and per h [4]. The chemical work done when 1 mol is transported from concentration C_1 to C_2 is $RT \ln C_1/C_2$. At $37°C$ the value of RT is 616 cal/mol, and the concentration term in the osmotic work therefore totals 7.5 cal/l cells per h. A further 0.3 cal/l per h has to be added for the electrical work involved in transporting 1 mmol of cations ($3Na^+ - 2K^+$) against the small resting membrane potential of the cells, making a total of 7.8 cal/l per h. Expressed in W, the rate of performance of osmotic work is 9 mW/l cells, so that if the Na^+ pump is assumed to have an overall efficiency of 25%, the energy cost of active transport is 36 mW/l cells. This is just under half the total heat output, which has been determined directly by Monti and Wadso [5] as 79 mW/l cells.

TABLE 1

PARTITION OF BASAL METABOLISM IN MAN

Organ	% body wt	Total energy consumption (W)	% basal metabolism	Energy for active transport (W)
Brain	2	16	18	12
Heart	0.4	8	9	0.5
Skeletal muscle	40	22	25	1
Liver	2	16	18	8
Digestive tract	5	19	22	15
Kidney	0.4	6	7	6
Skin and subcutaneous tissue	25	?	?	?
Total		87	99	42.5

There is, in this connexion, no such thing as a typical cell. An erythrocyte could, indeed, be regarded as rather atypical in being particularly impermeable to cations, requiring only about 1 (Na^+-K^+)-ATPase site in each μm^2 of membrane, as compared with 750 sites/μm^2 in a small non-myelinated nerve fibre [6]. But in compensation it is also atypical in being non-nucleated, so that its energy expenditure on turn-over of protein peptide bonds is unusually low. It may or may not be significant that old calculations of the osmotic work involved in maintaining the resting fluxes of labelled ions both in frog muscle [7] and in Sepia giant axons [8] also gave values in the region of 10% of the resting metabolism, as do more recent estimates for rabbit vagus nerve that I will not present in detail. In each case, therefore, the energy cost of active transport works out as around half the resting metabolism.

For some organs, calculation of osmotic work can yield quite misleading results. Thus much of the transport taking place in the kidney consists on the transfer of an isotonic solution from one compartment to another, which involves no net osmotic work. The apparent overall pumping efficiency of the kidney is consequently not much more than 1%. In the kidney of my standard man the total filtration rate for Na^+ is about 17 mmol/min, while the total O_2 consumption is 18 ml/min, so that the ratio of Na^+ transported to molecules of O_2 consumed in the process is about 20/1, rather than the limiting figure of 6/1 that would correspond to the hydrolysis of 1 ATP per Na^+. This indicates that the Na^+ pump does not necessarily operate in a strictly stoichiometric fashion, but can be geared up when the osmotic gradient is small.

In the last column of Table I, I have made some crude guesses at the energy specifically devoted to active transport. I shall not attempt to defend in detail my choices for the fraction of the total metabolism in the individual

158

organs assumed to be linked with transport processes, which range from 5% in skeletal and heart muscle to 100% in kidney. I will simply invite you to note that the total adds up to just about half the basal metabolism.

We can also approach the problem from the opposite direction, by estimating the magnitude of the energy expenditure on activities other than transport. In a man with a basal metabolism of 1800 kcal/day, protein synthesis in turn-over may be equivalent to the formation of as many as 4 mol peptide bonds with a total free energy of 16 kcal, and a total metabolic cost of 320 kcal (Blaxter, K.L., personal communication). Muscular contraction in circulation, gut movement and so on has been estimated from pressure-volume integrals to involve some 50 kcal of work; the consequent energy metabolism is say 300 kcal. The energy consumption for maintaining the resting tone of the skeletal musculature is probably about 270 kcal. The energy cost accounted for under these three headings is 890 kcal, or just under half the total metabolism.

It may be little more than chance that all these approaches agree in suggesting that roughly half the basal metabolism is expended on active transport. A rather crucial factor is the reliability of the estimates that have been made of the rate of turn-over of the proteins and other macromolecular constituents of the adult, and of the energy cost of resynthesis. However, even allowing for some uncertainty on this point, it would appear that active transport is the most important single energy consumer in man, and presumably in other higher animals as well. This supports the proposition that regulation of the ionic leakiness of cells may play a significant role in balancing supply and demand for energy in an adult. Dissipation of more energy by the Na^+ pump might be a convenient way of allowing the clutch to slip, as has been suggested for brown adipose tissue by Girardier et al. [9]. A subsidiary conclusion is that it might repay physiologists to look further into the problem of partitioning the energy expenditure between the different organs of the whole animal more precisely than seems to have been done so far. Using either direct or indirect calorimetry, it would not be technically difficult to improve considerably on the accuracy and meaningfulness of my Table I.

REFERENCES

1. Brožek, J. and Grande, F. (1955) Body composition and basal metabolism in man: correlation analysis versus physiological approach. Human Biol. 27, 22—31.
2. Durotoye, A.O. and Grayson, J. (1971) Heat production in the gastro-intestinal tract of the dog. J. Physiol. London 214, 417—426.
3. Webster, A.J.F., Osuji, P.O., White, F. and Ingram, J.F. (1974) Direct measurement of aerobic and anaerobic metabolism in the digestive tract of conscious sheep. In 6th Symp. Energy Metab. Farm Animals. E.A.A.P. Publ. No. 14. Universität Hohenheim Dokumentationsstelle: Stuttgart.

4. Whittam, R. (1964) Transport and diffusion in red blood cells. Arnold, London.
5. Monti, M. and Wadsö, I. (1973) Microcalorimetric measurements of heat production in human erythrocytes. I. Normal subjects and anemic patients. Scand. J. Clin. Lab. Invest. 32, 47—54.
6. Landowne, D. and Ritchie, J.M. (1970) The binding of tritiated ouabain to mammalian non-myelinated nerve fibres. J. Physiol. London 207, 529—537.
7. Keynes, R.D. and Maisel, G.W. (1954) The energy requirement for sodium extrusion from a frog muscle. Proc. Roy Soc. London Ser. B, 142, 383—392.
8. Hodgkin, A.L. and Keynes, R.D. (1954) Movements of cations during recovery in nerve. Symp. Soc. Exp. Biol. 8, 423—437.
9. Girardier, L., Seydoux, J. and Clausen, T. (1968) Membrane potential of brown adipose tissue. A suggested mechanism for the regulation of thermogenesis. J. Gen. Physiol. 52, 925—940.

Comparative Physiology — Functional Aspects of Structural Materials
Eds L. Bolis, H.P. Maddrell and K. Schmidt-Nielsen
© North-Holland Publishing Company — 1975 — Amsterdam

The energetics of ionic regulation

C.R. FLETCHER

Department of Pure and Applied Zoology, University of Leeds, Leeds LS2 9JT (U.K.)

ENERGETICS OF IONIC REGULATION

Osmotic and ionic regulation must inevitably consume some energy if the animal's body fluids are of different composition to its environment, and the exchanges of water and salt with the environment are not zero. This will constitute part of the resting metabolism, or basal metabolic rate, though if the exchanges with the environment increase as a result of activity or stress, as they appear to do in the teleosts [1,2] a part of the increase of metabolism under these conditions will be due to the increased energy requirements of osmotic and ionic regulation.

Investigators concerned with energetics will inevitably ask a number of questions, which may be formulated thus:

(i) What is the energy input required by the system?

(ii) What is the energy output (useful work) done by the system?

(iii) What is the efficiency of the system?

(iv) What are the component parts of the system, their efficiencies and interactions?

The overall effect of osmotic and ionic regulation is merely to maintain an internal steady-state, so the answer to questions (ii) and (iii) are each zero. Attempts to answer question (i) directly have usually been frustrated by the fact that osmotic and ionic regulation is rarely perfect, and the rates of many enzymes in the tissues change because of their ionic dependence [3]. That the substantial changes in metabolic rate of various animals when exposed to altered salinities is not principally due to the altered energy consumptions of ionic and osmotic regulation is clearly demonstrated by two facts: (a) that the metabolic rate of tissues not connected with ionic regulation changes markedly, and (b) that the animals which maintain the composition of their extracellular fluids most constant under changed external conditions show the smallest changes in metabolic rate [3].

Thus we are forced to reverse the usual procedure and attempt to answer question (iv) in the hope that by assembling the component parts we can answer question (i).

In broad outline, osmotic and ionic regulation in aquatic animals are quite well understood [4—6]. The freshwater animal gains water osmotically across its permeable epithelia, which are normally the respiratory surfaces. This, together with water taken in by feeding needs to be excreted as urine. Although freshwater animals usually produce a dilute urine this normally entails salt losses in excess of those gained by feeding, and salt has actively to be extracted from the surrounding water to replace the net loss. With the advent of radioactive tracers it has been possible to measure gross (i.e. unidirectional) fluxes of ions in the steady-state animal, and it is found that there is a gross efflux of ions across the respiratory surfaces normally rather larger than the losses in the urine. If such losses represent a passive diffusional gross efflux, the low salt concentration outside the animal will ensure that the corresponding gross passive influx will be negligible, and we therefore deduce that the freshwater animal needs actively to transport salt in at a rate several times the rate of urinary loss.

In the marine teleosts which hypoosmoregulate, water is lost by osmosis, and must be replaced by drinking the sea water and absorbing the water, which appears to be accomplished by absorbing the monovalent ions actively with water following osmotically. Thus the marine teleost needs actively to transport salt out of the body without significant amounts of water. The application of radioactive tracers has shown however that vastly more salt moves in across the body surfaces than does so by drinking. Experiment has shown that in many fishes a part of these ionic fluxes appear to be "exchange diffusion", defined as the obligatory exchange of one ion inside for an identical ion outside, and as such gives rise to no nett fluxes, but merely serve to confuse physiologists. Even when allowance is made for this however the passive diffusional fluxes appear to give rise to a nett influx of salt across the body surface, mainly the gills, which is several times the amount of salt drunk. Thus we deduce that the active transport of salt out has to balance the total of the amount drunk and the passive influx across the gills. I have calculated on this basis that in the marine teleosts *Serranus scriba* and *S. cabrilla* if salt transport were 100% efficient they would expend 16% of their basal metabolism on salt extrusion across the gills, even if neither ion passes through a significant potential "well" like Na^+ in a typical cell (Data published by Motais et al. [7] and others.)

However, it may be argued that the usual conceptual basis on which the gross ionic fluxes are analysed is inadequate and misleading, and the calculations referred to are spurious.

The assumptions that passive gross fluxes obey Ussing's flux ration equation [8]:

$$\frac{f_{0 \to 1}}{f_{1 \to 0}} = \frac{C_0}{C_1} \exp\left[-zF(\psi_1 - \psi_0)\right]$$

where f's are fluxes and C's are concentrations of an ion of charge z and ψ's are potentials, is based on the presumption that each ion moves through the membrane independently of the movement of any other ion of the same or different type and of the flow of solvent. The Goldman equations [9] which are also used (e.g. Smith [10]) depend on the additional premise that the electrical potential gradient through the epithelium is linear. Doubtful though these assumptions are, there are other indications that the conceptual framework is not entirely adequate, including the following:

(a) In the well investigated case of the crayfish it has been observed that the gross efflux of Na^+ varies according to the putative rate of active transport inwards [11], which Bryan explained by the hypothesis that the "pump" leaks and transports a Na^+ out for every 4 ions or so transported inwards.

(b) The rates of gross influx of ions across the gills of marine teleosts is higher by several orders of magnitude than the rates of branchial efflux in freshwater teleosts, and even when allowances are made for exchange diffusion and differences in concentration, the permeability, however, defined, of the marine teleost gill to ions is vastly greater than that of the freshwater teleost, and the same is true for euryhaline teleosts adapted to each salinity [5,6]. The necessity for such large permeability differences in similar or the same animals in different salinities is not accounted for by the conceptual framework outlined above.

(c) The low ionic fluxes of elasmobranchs which are ionically but not osmotically like the marine teleosts is also not accounted for by the present concepts.

All these observations become logically explicable if it is shown that the apparent passive permeabilities of these salt transporting epithelia is a direct consequence of their rate of salt transport, as may be demonstrated using the formalism of non-equilibrium thermodynamics. Firstly the phenomenological equations to describe active transport are set up, following the usual procedure [12], and then the energetics of such transport is examined as a function of the phenomenological coefficients to define optimum values for the latter.

A THERMODYNAMIC DESCRIPTION OF ACTIVE TRANSPORT

The basic statement of non-equilibrium thermodynamics is that the local rate of entropy production is everywhere greater than or equal to zero. To apply this principle to membrane phenomena it is convenient to sum the local rate of entropy production σ multiplied by the absolute temperature T per unit area of membrane. This is called the dissipation function Φ and can not be less than zero

$$\Phi = \int_0^{\Delta x} \sigma T \, \mathrm{d}x \geqslant 0$$

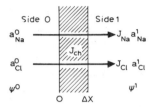

Fig. 1. Identification of symbols and directions.

where o and Δx denote the boundaries of the membrane. In an isothermal system in steady state, Φ may be equated to the sum of the products of each non-zero flow and its conjugate differences in electrochemical potential across the membrane. We consider here only three flows, of Na^+, Cl^-, and some chemical reaction coupled to the ionic flows, denoted respectively as J_{Na}, J_{Cl}, and J_{ch}; all other flows are assumed to be zero. The conjugate differences of electrochemical potential are $\Delta\tilde{\mu}_{Na}$, $\Delta\tilde{\mu}_{Cl}$, and the affinity A of the chemical reaction. J_{Na}, J_{Cl} are in equiv $\cdot$ m^{-2} $\cdot$ s^{-1}; $\Delta\tilde{\mu}_{Na}$, $\Delta\tilde{\mu}_{Cl}$ are in J/equiv and J_{ch} and A are defined such that their product is in J $\cdot$ m^{-2} $\cdot$ s^{-1}. Then:

$$\Phi = J_{Na}\Delta\tilde{\mu}_{Na} + J_{Cl}\Delta\tilde{\mu}_{Cl} + J_{ch}A$$

This may be rearranged by observing that in the absence of any external pathway through which an electrical current may flow, the ionic flows across the membrane must be electrically neutral, and hence $J_{Na} = J_{Cl} = J_s$ say. Also we may write expressions for $\Delta\tilde{\mu}_{Na}$ and $\Delta\tilde{\mu}_{Cl}$ under isothermal and isobaric conditions, viz.

$$\Delta\tilde{\mu}_{Na} = RT \ln\left[\frac{a^0_{Na}}{a^1_{Na}}\right] + F(\psi^0 - \psi^1)$$

$$\Delta\tilde{\mu}_{Cl} = RT \ln\left[\frac{a^0_{Cl}}{a^1_{Cl}}\right] - F(\psi^0 - \psi^1)$$

where the a's are ionic activities ψ's are electrical potentials, and R, T, F have their usual meanings.

Thus we may define a chemical potential difference between the salt solutions, $\Delta\mu_s$ without reference to electrical potentials:

$$\Delta\mu_s = \Delta\tilde{\mu}_{Na} + \Delta\tilde{\mu}_{Cl} = RT \ln\left[\frac{a^0_{Na} \cdot a^0_{Cl}}{a^1_{Na} \cdot a^1_{Cl}}\right]$$

and hence rewrite the dissipation equation:

$$\Phi = J_S \Delta \mu_S + J_{ch} A \tag{1}$$

Since the nett flow of ions across the epithelium must constitute a flow of neutral salt we may ignore its ionic nature. However, it is more usual to measure differences in osmotic pressure, $\Delta \pi_s$ than differences of chemical potential, so we replace $\Delta \mu_s$ by $\Delta \pi_s / \bar{c}_s$. This defines $\bar{c}_s$ as $\Delta \pi_s / \Delta \mu_s$, and it may be written as

$$\bar{c}_S = \frac{RT(a_{Na}^0 + a_{Cl}^0 - a_{Na}^1 - a_{Cl}^1)}{RT \ln (a_{Na}^0 \cdot a_{Cl}^0 / a_{Na}^1 \cdot a_{Cl}^1)}$$

Providing $a_{Na}^\circ / a_{Na}^1 \simeq a_{Cl}^\circ / a_{Cl}^1$, and a_{Na}° / a_{Na}^1 is not too different from unity it may be readily shown that

$$\bar{c}_S \simeq \tfrac{1}{4}(a_{Na}^0 + a_{Cl}^0 + a_{Na}^1 + a_{Cl}^1)$$

Thus the final form of the dissipation equation is

$$\Phi = J_S \Delta \Pi_S / \bar{c}s + J_{ch} A \tag{2}$$

We now proceed to set up phenomenological equations assuming that each flow (J_s and J_{ch}) is linearly related to both potential differences ($\Delta \pi_s / \bar{c}_s$ and A):

$$J_S = L_{11} \Delta \Pi_S / \bar{c}_S + L_{12} A \tag{3}$$

$$J_{ch} = L_{21} \Delta \Pi_S / \bar{c}_S + L_{22} A \tag{4}$$

The L's are phenomenological coefficients. The basic condition $\Phi \geqslant 0$ is satisfied for all A and $\Delta \pi_s / \bar{c}_s$ providing

$$L_{11} \geqslant 0, L_{22} \geqslant 0, \text{ and } L_{12} \cdot L_{21} \leqslant L_{11} \cdot L_{22}.$$

Also Onsager has shown theoretically that $L_{12} = L_{21}$. Thus within these constraints we may describe the active transport of salt.

In particular if we consider side "0" of the membrane to be the inside of a marine teleost, or the outside of a freshwater animal, $\Delta \pi_s < 0$ and the value of J_s required for salt balance is positive. Since $L_{11} \geqslant 0$ and $A > 0$ for an exothermic reaction, the required value of J_s can only be achieved with a sufficiently large and positive value of L_{12}.

It is pertinent to observe that as we vary $\Delta \pi_s$ both J_s and J_{ch} will vary.

However, it seems probable that in sufficiently brief experiments A, being the free energy of a chemical reaction will not alter significantly, though it will alter as the concentrations of reactants and products alter. Thus logically we identify the "permeability" of the membrane to salt ω_s at constant A:

$$\omega_s = \left.\frac{\partial J_S}{\partial \Pi_S}\right|_A = \frac{L_{11}}{c_S}$$

Thus $\quad L_{11} = \omega_S \bar{c}_S$ $\hfill$ (5)

Notice that ω_s is in no sense a leak, and represents the dependence of the rate of active transport on the gradient against which it operates, but it has the units of a permeability.

We may now proceed to analyse the energy input ϵ necessary to cause the required J_s

$$\epsilon = J_{ch} A.$$

Using (4) to substitute for J_{ch} and simplifying

$$\epsilon = A(L_{12}\Delta\Pi_S/\bar{c}_S + L_{22}A)$$

Substituting for A using Eqn (3) and simplifying,

$$\epsilon = \frac{L_{22}}{L_{12}^2} J_S^2 - \left(2\frac{L_{11}L_{22}}{L_{12}^2} - 1\right) J_S \Delta\Pi_S/\bar{c}_S + L_{11}\left(\frac{L_{11}L_{22}}{L_{12}^2} - 1\right)(\Delta\Pi_S/\bar{c}_S)^2 \quad (6)$$

Since we have identified $\Delta\pi_s < 0$ and $J_s > 0$, and have noted the constraints $L_{11} \geqslant 0$, $L_{22} \geqslant 0$, $L_{12}^2 \leqslant L_{11} \cdot L_{22}$ each term in Eqn (6) may be seen to be greater than or equal to zero. Thus it may be seen that ϵ is minimised by the conditions $L_{11} \to 0$, $L_{22} \to 0$ and $L_{12}^2 \to \infty$. However, the thermodynamic constraint $L_{12}^2 \leqslant L_{11} \cdot L_{22}$ limits the values that the phenomenological coefficients may assume. This may be expressed by writing

$$L_{12}^2 = \beta L_{11}L_{22} \quad \text{where} \quad 0 \leqslant \beta \leqslant 1.$$

Then Eqn (6) becomes

$$\epsilon = \frac{1}{\beta L_{11}} J_S^2 - \left(\frac{2}{\beta} - 1\right) J_S \Delta\Pi_S/\bar{c}_S + L_{11}\left(\frac{1}{\beta} - 1\right)(\Delta\Pi_S/\bar{c}_S)^2$$

We now see that ϵ is independent of L_{22} except that it is included in our definition of β. We may also use Eqn (5) to replace L_{11} by $\omega_s \bar{c}_s$

$$\epsilon = \frac{1}{\beta \omega_s c_s} J_s^2 - \left(\frac{2}{\beta} - 1\right) J_s \Delta\Pi_s / \bar{c}_s + \left(\frac{1}{\beta} - 1\right) \Delta\Pi_s^2 \omega_s / \bar{c}_s \qquad (7)$$

It may be observed that ϵ is minimised by the maximum possible value of β, which is the extent to which the coupling between the flow of salt J_s and the chemical reaction J_{ch} approaches its maximum possible value, $\beta = 1$. ϵ also depends on ω_s, and approaches infinity as ω_s tends either to zero or to infinity unless $\beta = 1$. Since ϵ is a continuous function of ω_s for all $\omega_s > 0$, it must have a definite minimum for some optimum value of ω_s, which may be identified by setting $\partial\epsilon/\partial\omega_s |_\beta = 0$ which gives the optimum value of ω_s as

$$\omega_{s,opt} = -J_s / \Delta\Pi_s \sqrt{1-\beta}$$

when ϵ is ϵ_{min}, given by

$$\epsilon_{min} = \frac{-J_s \Delta\Pi_s}{c_s} \cdot \frac{(1 + \sqrt{1-\beta})^2}{\beta}$$

As $\beta \to 1$ $\epsilon_{min} \to \epsilon_0 = -J_s \Delta\pi_s / \bar{c}_s$, the work which is actually being done.

ϵ has been calculated numerically for a range of values of β and ω_s by normalising ω_s to $-J_s / \Delta\pi_s$ and ϵ to ϵ_0, and the results are shown graphically in Fig. 2. The form of Eqn (7) may also be represented by writing ϵ_0/ϵ as a percentage efficiency of salt transport, and plotting isoefficiency lines in the plane defined by β and $-\omega_s \Delta\pi_s / J_s$ and these are shown in Fig. 3.

It is apparent that if active transport of salt is to occur with a reasonable efficiency the movement of salt and the driving chemical reaction must be very closely coupled ($\beta \to 1$) and the apparent permeability ω_s will be large ($\omega_s \geqslant -J_s / \Delta\pi_s$).

DISCUSSION

It seems logical that animals evolve in the direction of minimum dissipation, i.e. maximum efficiency, and this is in line with the physical principle that a stationary state is one of minimum dissipation [13]. In a constantly changing environment animals will never reach this point, but it seems probable that they will approach it quite closely. This provides the justification for assuming that animals may be expected to approach the conditions defined above. However other evolutionary pressures may also operate on some species. For example an euryhaline teleost which evolved a highly efficient system of salt transport in sea water ($\beta \to 1$, $\omega_s \to \infty$) would suffer a very

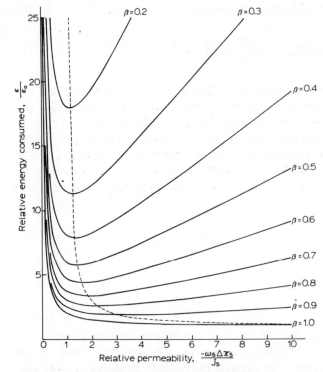

Fig. 2. Energy consumed relative to work done in causing a nett salt flow J_s against a concentration difference $\Delta\pi_s$, as a function of apparent permeability ω_s and coupling coefficient β.

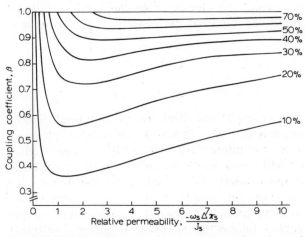

Fig. 3. Efficiency of active transport as a function of apparent permeability ω_s and coupling coefficient β.

high rate of salt loss on moving into fresh water until the necessary processes of adaptation to the freshwater environment were in progress, and thus natural selection might be expected to favour a value of ω_s below the optimal value for the value of β attained, to minimise such losses. However, as Figs 2 and 3 show, when $\omega_s < -J_s/\Delta\pi_s$ the process of active transport becomes energetically costly, and can not achieve efficiencies higher than 50%, and thus values of $\omega_s \ll -J_s/\Delta\pi_s$ are improbable.

The high value of the apparent permeability ω_s for maximum efficiency results in part from defining ω_s at constant A. If ω_s was defined at constant J_{ch} the optimum value would be $\omega_{s,opt} = -J_s\sqrt{1-\beta}/\Delta\pi_s$ which also approaches $-J_s/\Delta\pi_s$ when $\beta \ll 1$ but tends to zero as $\beta \to 1$. However as already pointed out measurement of changes in the nett flux J_s as a function of $\Delta\pi_s$ in short term experiments will approach the condition A = constant rather than J_{ch} = constant.

Rewriting the phenomenological equations (3) and (4) after substituting for L_{11} and L_{12}:

$$J_S = \omega_S\Delta\Pi_S + \sqrt{\beta\omega_S\bar{c}_S L_{22}} \cdot A \tag{8}$$

$$J_{ch} = \sqrt{\beta\omega_S L_{22}/\bar{c}_S} \cdot \Delta\Pi_S + L_{22}A \tag{9}$$

Now considering the significance of each term in Eqn 8, we can see that the first term has the form of a permeability or leak, and the second term corresponds to the part of the nett flux J_s which is driven by metabolism, i.e. the value J_s would assume if $\Delta\pi_s$ were reduced to zero. This term can not be greater than zero unless ω_s is also greater than zero.

These two terms would correspond to the conventional partition of the gross fluxes in marine teleosts to passive fluxes and to an active flux independent of $\Delta\pi_s$ (exchange diffusion giving rise to no nett fluxes). However it may be seen that the partition is misleading as the pair of equations describe active transport. The term $\omega_s\Delta\pi_s$ represents the changes in the rate of the pump as the chemical potential difference against which it operates is altered, and in no sense represents a real salt leak. Thus the partitioning of the isotopically measured fluxes as has been done in the past and assuming that that which appears to be a passive nett flux is a real leak which has to be compensated for by active transport (e.g. [14] is erroneous. This is not to argue that measurements of gross fluxes under various conditions is without value; indeed such observations together with measurements of gill potentials are probably the best way to unravel the mechanics of the pumps, and since non-equilibrium thermodynamics is not readily applicable to gross fluxes other conceptual bases will have to be used. However such isotopically measured fluxes need to be interpreted taking into consideration the fact that the rate of an efficient salt pump will vary markedly with the gradient against which it operates, and such changes may appear to be leaks, but are not themselves energetically dissipative.

To demonstrate that the conclusion reached that for efficient salt transport $\omega_s \geqslant -J_s/\Delta\pi_s$ is observed in practice we refer to the fishes. Unfortunately experiments in the past have not been designed to measure the variation of the nett flux J_s across the gills as a function of $\Delta\pi_s$ and we have to work from gross flux data, which will relate to the flounder, *Platichthys flesus* in sea water and fresh water. In sea water the flounder drinks, and presumably absorbs about 1 mequiv of Na^+ and 1.17 mequiv of Cl^- per kg/h [5]. From the work of Potts and Eddy [14] on the gill potentials and the nature of the ionic fluxes across the flounder gill they calculate that there appear to be passive permeabilities to Na^+ and Cl^- resulting in passive nett influxes across the gills of 1.59 mequiv Na^+ per kg/h and 1.61 mequiv Cl^- per kg/h [13]. (These figures are correct despite certain other errors in that paper.) Since Cl^- may be accompanied by cations other than Na^+, we will equate the salt fluxes to Cl^- fluxes. We may thus identify $J_s = 1.17$ mM salt per kg/h and $\omega_s \Delta\pi_s = -1.61$ mmol salt per kg/h. Thus

$$\omega_s = \frac{-1.61\,J_S}{1.17\,\Delta\Pi_S} = -1.376\,J_S/\Delta\Pi_S$$

This is within the expected range, and would be the optimum ω_s for a β of 0.5, when the efficiency of salt transport would be 17%. But if β were as high as 0.9 ϵ would still be only 10% above the minimum for this value of β. From Fig. 2 it can however be seen that the efficiency of active transport can not be higher than 57% with this value of ω_s.

For the flounder in fresh water, data seems to be available only for Na^+ fluxes, and hence it is assumed that the nett fluxes of Cl^- are similar to those of Na^+. In flounders acclimatised to 0.2 equiv Na^+/l the urinary Na^+ losses are about 20 μequiv per kg/h whilst the gross efflux of Na^+ across the gills is about 40 μequiv per kg/h [5]. Assuming the latter behaves as a passive efflux, the passive influx will be negligible and we may equate the passive efflux to $-\omega_s \Delta\pi_s$. Thus $J_s = 20$ μequiv per kg/h and $\omega_s \Delta\pi_s = -40$ μequiv per kg/h, and hence

$$\omega_s = \frac{-40\,J_S}{20\,\Delta\Pi_S} = 2J_s \Delta\Pi_s$$

This is again well within the expected range, and would correspond to $\omega_{s,opt}$ if $\beta = 0.75$, when salt transport would be 33% efficient, and with no value of β could its efficiency exceed 65%.

The analysis presented here assumes that the flows are linearly related to the driving forces, an assumption which may theoretically be justified only close to thermodynamic equilibrium. If, however, salt transport is to operate efficiently this assumption may be expected to be reasonable. Large degrees of non-linearity may be expected as a result of large deviations of the external conditions from those to which the animal was acclimatised. The analysis is also incomplete insofar as it ignores the movement of other ions and water, and the possibility of interactions between these flows and flow of salt. A more general analysis including the probability of solute/solvent interactions gives some other interesting conclusions, but upholds the conclusions reached here that insofar as animals evolve towards minimising their energy expenditure on salt and water regulation the coupling coefficient β will be high, and ω_s will be greater than or approximately equal to $-J_s/\Delta\pi_s$.

Thus we may move closer to understanding the properties and energetics of ionic transport across the transporting epithelia of aquatic animals. Until, however, experiments can be devised to measure β, for example by measuring J_{ch} as $\Delta\pi_s$ is varied we shall not be able to estimate accurately the efficiency of salt transport and the energy consumed by it. Kedem [16] has suggested that J_{ch} might be identified with the oxygen consumption of the tissue, but this presents practical difficulties in vivo as the transporting epithelium is also the tissue responsible for most of the animal's respiratory gas exchange, and only a small proportion of the oxygen passing into the gill will not leave the gill at the serosal side. The attention currently being paid to perfused gill preparations in various laboratories may well result in suitable preparations being available to make such measurements in vitro, though the problems of relating behaviour in vitro to behaviour in vivo will then be raised.

REFERENCES

1. Mayer, N. and Maetz, J. (1967) Axe hypophyso-interrénalien et osmorégulation chez l'Anguille en eau de mer. Etude de l'animal intact avec notes sur les perturbations de l'équilibre minéral produites par l'effet de "choc". C.R. Hebd. Séance Acad. Sci. Paris 264D, 1632—1635.
2. Randall, D.J., Baumgarten, D. and Malyusz, M. (1972) The relationship between gas and ion transfer across the gills of fish. Comp. Biochem. Physiol. 41 A, 629—637.
3. Schoffeniels, E. (1973) Amino acid and Cellular Volume Regulation. In Comparative Physiology — Locomotion, Respiration, Transport, and Blood. (Bolis, L., Schmidt-Nielsen, K. and Maddrell, S.H.P., eds), North Holland, Amsterdam and Elsevier, New York.
4. Potts, W.T.W. and Parry, G. (1964) Osmotic and Ionic Regulation in Animals. Pergamon Press, London and New York.
5. Maetz, J. (1971) Fish gills: mechanisms of salt transfer in fresh water and sea water. Phil. Trans. Roy. Soc. Lond. Ser. B. 262, 209—249.
6. Motais, R. and Garcia-Romeu, F. (1972) Transport mechanisms in the teleost gill and the amphibian skin. Annu. Rev. Physiol. 34, 141—176.
7. Motais, R., Garcia-Romeu, F. and Maetz, J. (1965) Mécanisme de l'euryhalinité. Etude comparée de Flet (euryhalin) et du Serran (sténohalin) au cours du transfert en eau douce. C.R. Hebd. Séance Acad. Sci. Paris 261, 801—804.
8. Ussing, H.H. (1949) The distinction by means of tracers between active transport and diffusion. Acta Physiol. Scand. 19, 43—56.

9. Goldman, D.E. (1943) Potential, impedence and rectification in membranes. J. Gen. Physiol. 27, 37—60.
10. Smith, P.G. (1969) The ionic relations of *Artemia salina* (L). II. Fluxes of sodium, chloride, and water. J. Exp. Biol. 51, 739—757.
11. Bryan, G.W. (1960) Sodium regulation in *Astacus fluviatilis*. II. Experiments with sodium depleted animals. J. Exp. Biol. 37, 100—112.
12. Katchalsky, A. and Curran, P.F. (1967) Non-equilibrium Thermodynamics in Biophysics. Harvard U.P. Cambridge.
13. Prigogine, I. and Wiame, J.M. (1946) Biologie et Thermodynamique des phénomènes irréversibles. Experientia 2, 451—453.
14. Potts, W.T.W., Fletcher, C.R. and Eddy, B. (1973) An analysis of the sodium and chloride fluxes in the flounder *Platichthys flesus*. J. Comp. Physiol. 87, 21—28.
15. Potts, W.T.W. and Eddy, B. (1973) Gill potentials and sodium fluxes in the flounder *Platichthys flesus*. J. Comp. Physiol. 87, 29—48.
16. Kedem, O. (1961) Criteria of Active Transport, in Membrane Transport and Metabolism. (Kleinzeller, A. and Kotyk, A., eds), pp. 87—93, Academic Press, London and New York.

Comparative Physiology — Functional Aspects of Structural Materials
Eds L. Bolis, H.P. Maddrell and K. Schmidt-Nielsen
© North-Holland Publishing Company — 1975 — Amsterdam

Biochemical energetics for fast and slow muscles

G. GOLDSPINK

Muscle Research Unit, Department of Zoology, University of Hull (U.K.)

INTRODUCTION

The energy demand of muscle as a tissue is very variable because it depends on the level of activity of the tissue. Millikan [1] followed the rate of utilization of O_2 used by muscle by measuring the rate at which myoglobin was deoxygenated in situ. He concluded that resting muscle uses 0.04—0.06 cm^3/min per kg tissue whilst tetanised muscle used up to 3.1 cm^3/min per kg tissue. Thus there is 50—80-fold increase in the energy demand from rest to full activity. Under certain circumstances the increase in energy demand may be even greater, for example, a muscle doing a short burst of maximal isotonic work.

For many years now physiologists have been interested in energy demands of muscle and in particular, the efficiency of muscle as a machine for converting energy into work. Earlier measurements of energy utilization by muscles were carried out using either O_2 consumption or heat measurements. These measurements have provided some very useful data, for example, Hill [2] estimated that when an athletic man pedals a bicycle ergometer at a high rate of working the muscles involved in this exercise have an overall efficiency of nearly 30%. This efficiency value of course includes the efficiency of such processes of glycolysis, oxidative phosphorylation and ion pumping as well as efficiency of the contractile process itself. However, when one is measuring the thermodynamic efficiency of the contractile process, heat and work measurements are not sufficient: chemical measurements are essential [3]. Over the past few years we have studied the efficiency of the contractile process in different kinds of muscles using biochemical methods. What has emerged from our work is that the energy demand of muscle as a tissue is not only dependent on the level of activity but also dependent on the nature of the contractile activity and the kind of muscle fibres that are involved.

RESULTS

Fast and Slow Muscles

The maximum tension that can be generated by different striated muscles ranges from 0.5 to 4 kg/cm^2 [4]; however, this range can be probably reduced if the tension is corrected for the differences in myofibril density of the different muscles. The range in the maximum rate of shortening of muscles throughout the animal kingdom is much greater. Differences in the rate of shortening are not only apparent between muscles from different species but also between the muscles within the same animal. In the mammal most muscles are slow at birth, however, as growth progresses some muscles develop into fast muscles whilst others remain slow [5,6]. This differentiation into fast and slow muscles has been shown to be dependent on the type of innervation the muscle fibres receive [7].

The biochemical differences between muscles of different speeds of contraction have been studied by Bárány [8] who showed that the rate of shortening was generally proportional to the activity of the ATPase of the myosin prepared from the muscles. The post-embryonic differentiation into fast and slow muscles is a very interesting problem. The reason why some muscles begin to produce a predominance of "fast myosin" whilst others continue to synthesise "slow myosin" has not yet been answered in molecular terms. However, on examining the subject of fast and slow muscle it was realised that no satisfactory reason had been given for the existence of fast and slow muscles — what possible advantage could it be to an animal to have slow muscles?

There is a partial explanation for the slower muscles of larger mammals which was presented by Hill [9]. The overall rate of contraction of a muscle is dependent on rate of shortening of each sarcomere (this is dependent on the rate at which the myosin ATPase works and is called the intrinsic velocity), but it is also dependent on the number of sarcomeres in series. Hill argued that the very long muscles of the larger mammals would develop force at such a rate that their tendons would be pulled off the bones unless, of course, these muscles have a slower intrinsic speed of shortening. This argument, I believe, still holds good and if one looks at the fibre composition of mammals of different sizes one finds that in general, the larger the animal, the greater the percentage of slow fibres in the muscle. However, this does not explain why there should be muscles with different intrinsic rates of shortening within the same animal. We felt that the reason for the existence of fast and slow muscles must be really a question of efficiency. The slow muscles must in fact be able to perform some types of activity more efficiently than the fast muscles. The ideal opportunity arose to test this hypothesis when I spent a sabbatical year working with R.E. Davies at the University of Pennsylvania.

Biochemical Methods for Measuring Energy Consumption During Muscular Contraction

The important reactions in the immediate energy supply of muscle are shown in Fig. 1. Early attempts to measure the usage of high energy phosphate during muscular contraction failed because the ATP that was hydrolysed in the muscles was immediately rephosphorylated by the creatine phosphate via the creatine phosphokinase reaction. In mammalian muscle the creatine phosphate levels are 3--4 times higher than the ATP levels and therefore no change can be detected in the ATP levels of untreated muscles except during very brief contractions. However, in 1962 Cain and Davies [10] were successful in measuring ATP breakdown during contraction by blocking the creatine phosphokinase reaction using 1-fluoro-2,4-dinitrobenzene (FDNB). This is a fairly non specific inhibitor which blocks most of the metabolic reactions but, fortunately, it does not block the myosin ATPase reaction.

An alternative approach was used by Mommaerts et al. [11] who prevented restitution of the creatine phosphate levels by blocking glycolysis with iodoacetate and oxidative phosphorylation by bubbling N_2 through the Ringer solution. With this latter method the high energy phosphate utilization is measured as the decrease in the creatine phosphate level or the increase in the free creatine level.

In our experiments we have used both of these methods. However, we usually favour the latter method because iodoacetate/nitrogen-treated muscles have more high energy phosphate available (ATP and creatine phosphate) and therefore it is possible to get them to do more work.

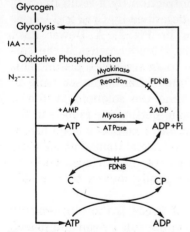

Fig. 1. The main reactions involved in the immediate energy supply of muscle. The two methods of preventing the replenishment of the immediate supply are shown. These involve the use of fluoro-2,4-dinitrobenzene (FDNB) or the use of iodoacetate (IAA) and nitrogen.

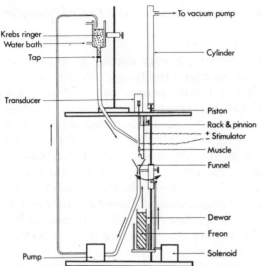

Fig. 2. Apparatus used for rapid freezing of muscles during contraction. When the solenoid switch is activated the Dewar containing Freon cooled to $-160°C$ is pulled up by the evacuated cylinder, immersing the muscle in the Freon.

Pairs of muscles are incubated for 35 min in a solution of the inhibitor in a suitable Ringer solution maintained at $4°C$ and through which N_2 is bubbled. For some fast muscles we have found it desirable to use cyanide as well as iodoacetate and nitrogen and to shorten the incubation period. One of the pair of poisoned muscles is then stimulated whilst the contralateral muscle acts as the control.

The stimulated muscles are frozen during contraction by a rapid freezing method which is similar in design to that first described by Cain and Davies [20]. A diagram of the apparatus is shown in Fig. 2. A Dewar flask containing cold Freon at $-160°C$ is placed in a holder beneath the suspended muscle. The Dewar holder is connected to an evacuated piston. This holder is restrained with a solenoid arm. The time of freezing is set using an electronic timing device and at the desired time the solenoid switch is activated and the Dewar flask is pulled up, thus immersing the muscle in the cold Freon. The muscle freezes in about 100 ms and hence its metabolism is abruptly arrested. During the contraction the tissue is maintained at $35°C$ by continually dripping the warm Ringer solution over its surface. The collecting vessel for the Ringer solution is automatically moved to one side when the Dewar flask is pulled up by the piston.

After a muscle has been frozen, the Dewar of Freon is then lowered by releasing the vacuum. The muscle is cut down whilst still frozen and placed in a pre-cooled stainless-steel tube. It is then pulverised in this tube and extracted with 0.5 M $HClO_4$ at $0°C$. Following centrifugation at $0°C$, aliquots of the supernatant are assayed for total and free creatine [12].

Cost of Development and Maintenance of Isometric Tension by Mammalian Fast and Slow Muscle

As the slow muscles are usually associated with a postural function, it seemed worthwhile to look at the cost of developing and maintaining isometric tension for the hamster biceps brachii (fast) and the soleus (slow) muscles. These muscles were stimulated whilst being held at a fixed length and the tension developed was related to the amount of high-energy phosphate used. This was done for contractions of 1.2-s, 30-s and 60-s duration. The results are summarised in Table I and are from Awan and Goldspink [13]. From the table it will be seen that much of the utilization of high-energy phosphate occurs during the first second or so during which time the muscles are developing isometric tension. After this initial period the muscles are merely maintaining tension, hence this is associated with a lower rate of expenditure of high-energy phosphate. In order to compare the cost of developing and maintaining isometric tension by the two types of muscle, the area under the contraction traces was measured and expressed as gramme seconds (g · s). The relative cost (economy) was then expressed as the number of gramme seconds that could be developed and maintained per μmol of high energy phosphate. As will be seen from the table, the slow soleus muscle was less costly than the fast biceps brachii and this was more marked in the longer contractions. For contractions of 60-s duration the slow muscle was about 2.5 times less costly than the fast muscle.

TABLE I

THE EXPENDITURE AND COST OF DEVELOPING AND MAINTAINING ISOMETRIC TENSION BY THE HAMSTER BICEPS BRACHII (FAST) AND SOLEUS (SLOW) MUSCLES

The amount of tension developed and maintained is measured by integrating the area under the tension curves and is expressed as gramme seconds (g · s)

	Length of contraction	1.2 s	30 s	60 s	
Expenditure (μmol high-energy phosphate/g)	Fast	1.5 ± 0.06	2.2 ± 0.05	2.5 ± 0.05	In vitro
	Slow	1.1 ± 0.03	1.5 ± 0.02	1.8 ± 0.01	
	Fast	—	—	2.2 ± 0.01	In vivo
	Slow	—	—	1.1 ± 0.01	
Cost (g · s)/μmol high-energy phosphate)	Fast	99 ± 7	2483 ± 110	3635 ± 180	In vitro
	Slow	143 ± 8	6377 ± 250	9012 ± 230	
	Fast	—	—	4997 ± 431	In vivo
	Slow	—	—	9963 ± 560	

Also given in this table are some measurements for the energy expenditure of muscles in vivo. These were made on muscles of hamsters in which one tendon was dissected free and connected to the transducer. These in vivo results are not of course the same as those in which the muscles were subjected to iodoacetate and nitrogen inhibition because the in vivo muscles were able, to some extent, to replenish their high-energy phosphate levels. The ability to replenish was seen to be greater in the slow muscle. This is what one might expect because slow fibres tend to have more mitochondria and much higher levels of oxidative enzymes than fast fibres.

From this set of experiments it seems therefore that mammalian slow muscles are adapted to maintaining isometric tension, and this is presumably why they are usually found to have a postural rôle.

Reason for the Utilization of Energy During Isometric Contractions

When a muscle is contracting isometrically it is in fact doing no external work because in order to do work it is necessary for the muscle to lift a certain weight through a given distance. However, the muscle must be performing some kind of internal work. In order to find out about the nature of this internal work we [14] measured the movement of sarcomeres by a laser beam diffraction method. The sarcomeres are the repeated units along the myofibrils in the muscle fibre and these give strong diffraction lines when a suitable light source is used such as a helium/neon low power laser. In one set of experiments the diffraction lines were filmed at 60 frames/s before and during isometric contractions of the chicken posterior latissimus dorsi muscle. The film was then analysed frame by frame using a densitometer and the distance between the density peaks of the diffraction lines was measured.

When stimulation is commenced there is initial shortening of the sarcomeres as the muscle develops isometric tension. This is no doubt due to the compliance of the tendons and recording apparatus. However, following this, the sarcomeres in the region of the beam begin to fluctuate in length. This is true even when the muscle is stimulated at high frequencies. (The frequency of fluctuation is not in fact synchronous with the frequency except at low frequencies.) It seems therefore that even during isometric contraction the sarcomeres are fluctuating in length and that this is one of the sources of internal work. Experiments on slow muscles indicate that the frequency of fluctuation is considerably less than for the fast muscle. The reason why the sarcomeres fluctuate in length up to 90 nm is not known although it may be due to a slightly erratic ATP supply causing slipping of the filaments in one part of the muscle. However it does indicate that cross-bridges are active even when the muscle as a whole is not changing length. More direct evidence for cross-bridge activity is obtained from the X-ray diffraction data presented by Haselgrove [15] in the contractile systems section of this book (p. 127). In order to explain the energy demands of different types of muscle we have of course to think in terms of the cross-bridge activity of these muscles when engaged in different kinds of contraction.

Cost of Producing Isometric Tension for Some Other Types of Vertebrate Muscle

Vertebrate striated muscle fibres may be classified into two major types; true slow or tonic fibres and twitch or phasic fibres. The hamster muscles used for the above studies are both examples of twitch muscles. The biceps brachii is a fast twitch muscle and the soleus is a slow twitch muscle. Therefore it was felt that this biochemical approach to measuring the energy demands during contraction should be extended to other kinds of vertebrate muscle. A good example of a true slow muscle is the chicken anterior latissimus dorsi (ALD) which runs from the spinal processes of the vertebral column to the humerus of the wing. The adjacent muscle to the ALD is the posterior latissimus dorsi (PLD) which is a pure fast twitch muscle and this therefore provides a very convenient comparison muscle.

Values for the cost of isometric tension are given in Table II together with some information on the speed of contraction and the rate of the myosin ATPase of these muscles and the hamster muscles.

From Table II it will be seen that the chicken ALD can produce and maintain a very considerable amount of isometric tension per μmol of high energy phosphate. In this respect it is 15—18 times less costly than the

TABLE II

SUMMARY OF MEASUREMENTS OF THE COST OF PRODUCING ISOMETRIC TENSION FOR DIFFERENT KINDS OF VERTEBRATE MUSCLE

Muscle	Cost of isometric tension for 1 min contraction (g · sec μmol high-energy phosphate)	Maximal velocity of shortening lengths s^{-1}	Myosin ATPhase (μmol P_i mg^{-1} · min^{-1})
Hamster biceps brachii (fast twitch)	3 635	8.0	0.8
Hamster soleus (slow twitch)	9 021	2.0	0.38
Chicken ALD (true slow)	54 200	1.0	0.2
Chicken PLD (fast twitch)	200*	10.0	—
Tortoise retractor femoris	7 300†	—	—
Tortoise penis retractor	15 087†	—	—
Tortoise neck retractor	14 503†	—	—
Frog sartorius	935†	—	—

* for a 10 s duration
† measurments carried out at 5°C

mammalian fast twitch muscle, about 6 times less costly than the slow twitch muscle and many times more efficient than its counterpart, the chicken PLD. The chicken PLD, in fact, was so costly and fatigued so quickly that the longest period over which the measurements could be made was 10 s. The role of the ALD muscle is to hold the wings back against the body and therefore it has to remain contracted for long periods of the time. From an energetic point of view, it is obviously well adapted to this function. However, the isotonic efficiency of the ALD is low (160 g · cm · μmole P^{-1} which is equal to an efficiency of 38%) as compared with most other muscles which have an efficiency of between 60% and 70%, therefore it is just possible that the contractile mechanism involved is slightly different in these true slow muscle fibres. It may well involve some sort of catch mechanism similar to that of the molluscan anterior byssus retractor muscle which has also been shown to use very little energy during the maintenance of tension [16]. The role of the PLD, one assumes, is just that of pulling the wings back so that the ALD can continue to hold them in place.

From Table II we also see that tortoise muscles, especially the penis and neck retractor muscles, are able to maintain a lot of isometric tension per μmol of ATP. These muscles are still more costly than the chicken ALD, however they are many times less costly than other reptilian or amphibian muscles, e.g., the frog sartorius muscle. The tortoise has, of course, a rather special postural problem in that it carries a heavy shell and therefore it has, by necessity, to have slow muscles. The penis and neck retractor muscles are muscles which also have to remain contracted for long periods of time.

Efficiency of Mammalian Fast and Slow Muscles in Doing Work

In order to find the maximum efficiency of fast and slow muscles in performing work, they were made to contract isotonically when loaded with different weights. The amount of creatine phosphate hydrolysed was measured and the isotonic efficiency calculated as the work done per μmol of high energy phosphate. In each case the muscles were made to do several contractions so that the work done was determined by multiplying the sum total of the distance shortened by each muscle by its load. As the efficiency of the muscle is dependent on the rate at which it is made to shorten rather than its load, the efficiency values were plotted against the measured rate of shortening for each individual muscle. Values for the hamster fast and slow twitch muscles are shown in Fig. 3. From this figure it will be seen that the slow muscle is working most efficiently when it is contracting at about 1 muscle length per second and the fast muscle is working most efficiently at 5 muscle lengths per s. From the data given in Fig. 3 it can be seen that for slow isotonic contractions it is energetically advantageous to use slow muscle fibres. This is in accordance with the recent findings of Gollnick et al. [21,22], who looked at glycogen depletion in different types of fibres in

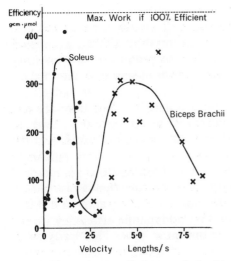

Fig. 3. The isotonic efficiency of the biceps brachii (×) and soleus (●) muscles plotted against the rate of shortening of each muscle. The different rates of shortening in muscle lengths per s were obtained by loading the muscles with different sized weights.

human muscle, during isometric and isotonic exercise. These workers found that the slow fibres were recruited first in both kinds of exercise. In the isometric exercise the fast fibres were not recruited until the tension required was greater than that which could be produced by using the slow fibres alone. The isotonic exercise consisted of peddling a bicycle ergometer at different speeds. It was found that at slower rates of peddling only the slow fibres were used and as the rate of peddling increased the fast fibres began to be recruited. It seems therefore that slow twitch fibres are not only used for isometric contractions of a postural nature but also for slow isotonic movement.

DISCUSSION

Over the course of evolution individual muscles have become adapted for the particular role they have to play. The energetic efficiency of a muscle in carrying out a certain type of contractile activity seems to have been one of the main factors that has been selected for during evolution and sometimes this has apparently been at the expense of other desirable characteristics. This is understandable because the struggle for survival is essentially one of obtaining sufficient energy, in the form of food, and the species that have been able to obtain and use this energy most efficiently have been those that have survived. As muscle is a tissue that uses large quantities of energy there has obviously been considerable selection pressure for efficient muscle. For this reason muscles that have a postural role have evolved with a slow myosin ATPase as this enables them to produce isometric contractions at less cost than would be the case if a muscle with a fast myosin ATPase were used. Fast muscles on the other hand are well suited to producing large isotonic movements because their high speed of contraction means that they have a high work output (power).

Slow muscles are not only used for postural activities but they are also found in situations where slow isotonic contractions are desirable. Two examples are, the red muscle in fish which is used for slow cruising, and the smooth muscle of the gut which is required to produce slow peristaltic waves. One can see from Fig. 3 that if slow isotonic contractions are required it is energetically advantageous to use a slow muscle rather than a fast muscle in which only a few, hence heavily loaded, motor units are activated at any one time. In order to obtain the maximum thermodynamic efficiency from a given muscle it must be shortening at its optimum rate. This leads us to consider what is the highest possible efficiency at which a muscle can work. Kushmerick and Davies [17] found that the efficiency of frog sartorius muscle working at optimum velocity was over 66% and if this was corrected for the energy used for calcium pumping the efficiency was found to be 98 ±15%.

Some slow muscles have a slightly higher maximum efficiency for isotonic contraction including ion pumping. For example, the hamster soleus is able to do 320 g · cm of work per μmole P and the tortoise retractor femoris is able to do 350 g · cm of work per μmol P out of the possible 450 g · cm of work that is theoretically available from each high energy phosphate derived from creatine phosphate. These muscles are no doubt able to work at these higher efficiencies of 70—80% because there is less ion pumping associated with their activation. However if we take 66% for the frog sartorius as representing muscles in general, and correct this for the efficiency of obtaining energy from glycolysis and oxidative phosphorylation which are approximately 50% efficient, we have a value of just over 30% which compares favourably with that obtained by A.V. Hill [2] for muscles in situ, as mentioned in the introduction.

Most vertebrate muscles are not composed entirely of fast muscle fibres or entirely of slow muscle fibres, they do in fact contain a mixture of fast, slow and intermediate motor units. These can be detected physiologically and histochemically [18]. The reason for having different kinds of fibres within a muscle can be appreciated if we consider Fig. 3. This shows a peak of efficiency for the slow fibres (of the soleus) and a peak of efficiency for the fast fibres (of the biceps brachii) and these peaks are quire widely separated. The presence of different types of fibres within a muscle thus enables the muscle to be versatile, in other words to maintain posture and also to shorten reasonable efficiently over the range of speeds at which the animal moves. However, because the efficiency peaks of the fast and slow fibres are widely separated it is likely that even mixed muscles will have only two rates of shortening at which they can work with a high degree of efficiency.

The development of force by muscle is believed to be due to the interaction of the myosin cross-bridges with the actin filaments. In order to understand the rate of energy utilization by the muscles during different kinds of contractile activity, it is helpful to speculate on the way in which the cross-bridges might work. ATP is believed to combine with the cross-bridge before it engages with the actin filament. In measuring the high energy phosphate utilization therefore one is really measuring the rate at which the cross-bridges are being reprimed.

The longer cycle time of the slow muscle cross-bridges presumably means that the cross-bridges remain attached to the actin for a longer time and, within a given period, fewer of them will need to be reprimed. The slow muscle can therefore produce isometric tension at very little cost. However, if one considers the situation when the muscle is required to contract isotonically it will be appreciated that a slow cross-bridge cycle may not be an advantage because the cross-bridges that are engaged and shortened will be working against those that are just holding or at least doing very little shortening as seems to be the case for the chicken ALD which has a relatively low isotonic efficiency ($160 \text{ g} \cdot \text{cm} \cdot \mu\text{mol } P^{-1}$).

The situation where a fast muscle is required to contract very slowly is perhaps more readily understood. For the cross-bridges to make the best use of each molecule of ATP it is necessary for them to move the actin filaments through a given distance before they have to be reprimed. When a fast muscle is made to contract slowly then the actin filaments will only have moved a fraction of this distance within the cross-bridge cycle time and consequently the efficiency at which the muscle is working will be very much reduced. This effect would, of course, be far less marked in a slow muscle because of its long cycle time. This argument can be extended to the situation where the slow muscle is required to shorten faster than its optimum speed; in this case the cross-bridge will not detach quickly enough and therefore we have the effect already mentioned, where some of the cross-bridges will be working against those that are holding.

The energetics of different kinds of muscle is a fascinating subject to study because it not only provides answers regarding the evolution of the muscles but it also allows one to speculate about the molecular events that are involved in the contractile process. However, we will have to await the elucidation of the fundamental contractile events before we know whether our speculations are correct and before we can explain fully the exact nature of the energy demand the different kinds of muscle.

ACKNOWLEDGEMENT

This work was carried out whilst the author was in receipt of research grants from the Muscular Dystrophy Associations of the United States.

REFERENCES

1. Millikan, G.A. (1937) Experiments on muscle haemoglobin in vivo; the instantaneous measurement of muscle metabolism. Proc. Roy. Soc. London Ser. B 123, 218—241.
2. Hill, A.V. (1934) The efficiency of bicycle pedalling. J. Physiol. London 82, 207—210.
3. Wilkie, D.R. (1960) Thermodynamics and the interpretation of biological heat measurements. Progr. Biophys. Biophys. Chem. 10, 260—298.
4. Wilkie, D.R. (1954) Facts and theories about muscle. Progr. Biophys. 4, 288—324.
5. Buller, A.J., Eccles, J.C. and Eccles, R.M. (1960) Differentiation of fast and slow muscles in the cat hind limb. J. Physiol. London 150, 399—416.
6. Close, R. (1964) Dynamic properties of fast and slow skeletal muscles of the rat during development. J. Physiol. London 173, 74—95.
7. Buller, A.J., Eccles, J.C. and Eccles, R.M. (1960) Interaction between motoneurons and muscles in respect of the characteristic speeds of their responses. J. Physiol. London 150, 417—438.
8. Bárány, M. (1967) ATPase activity of myosin correlated with speed of muscle shortening. J. gen. Physiol. 50, 197—218.
9. Hill, A.V. (1950) The dimensions of animals and their muscular dynamics. Proc. Roy. Instr. 34, 450—473.
10. Cain, D.F. and Davies, R.E. (1962) Breakdown of adenosine triphosphate during a single contraction of working muscle. Biochem. Biophys. Res. Commun. 8, 361—366.
11. Mommaerts, W.F.H.M., Seradarian, K. and Maréchal, G. (1962) Work and chemical change in isotonic muscular contractions. Biochim. Biophys. Acta 57, 1—12.
12. Eggleton, P., Elsdon, S.R. and Cough, N. (1943) The estimation of creatine with diacetyl. Biochem. J. 37, 526—529.
13. Awan, M.Z. and Goldspink, G. (1972) Energetics of the development and maintenance of isometric tension by mammalian fast and slow muscles. J. Mechanochem. Cell Motility 1, 97—108.
14. Goldspink, G., Larson, R.E. and Davies, R.E. (1970) Fluctuations in sarcomere length in the chick anterior latissimus dorsi muscle during isometric contraction. Experientia 26, 16—18.
15. Hasselgrove, J.C. (1975) This volume, pp. 127—138.
16. Nauss, K.M. and Davies, R.E. (1966) Changes in inorganic phosphate and arginine during the development, maintenance and loss of tension in the anterior byssus retractor muscle of mytilus edulis. Biochim. Z. 345, 173—187.
17. Kushmerick, M.J. and Davies, R.E. (1969) The chemical energetics of muscle contraction II. The chemistry, efficiency and power of maximally working sartorius muscles. Proc. Roy. Soc. Lond. Ser. B 174, 315—353.
18. Close, R. (1972) Dynamic properties of mammalian skeletal muscles. Physiol. Rev. 52, 129—197.
19. Awan, M.Z., Frearson, N., Goldspink, G. and Waterson, S.E. (1972) Biochemical efficiency of smooth muscle and different types of striated muscle. J. Mechanochem. Cell Motility 1, 225—232.

20. Cain, D.F. and Davies, R.E. (1964) Rapid arrest of metabolism with melting Freon. In Rapid mixing and sampling techniques in biochemistry. (Chance, B., Eisenhardt, R.H., Gibson, Q.H. and Lowberg-Holm, K.K.), pp. 229—237, Academic Press, New York.
21. Gollnick, P.D., Piehl, K. and Saltin, B. (1974) Selective glycogen depletion patterns in human muscle fibres after exercise of varying intensity and at varying pedalling rates. J. Physiol. 241, 45—58.
22. Gollnick, P.D., Karlsson, J., Piehl, K. and Saltin, B. (1974) Selective glycogen depletion in skeletal muscle fibres of man following sustained contractions. J. Physiol. 241, 59—68.

Comparative Physiology — Functional Aspects of Structural Materials
Eds L. Bolis, H.P. Maddrell and K. Schmidt-Nielsen
© North-Holland Publishing Company — 1975 — Amsterdam

Bioenergetics of teleost fishes: Environmental influences

F.W.H. BEAMISH, A.J. NIIMI and P.F.K.P. LETT

Department of Zoology, University of Guelph, Guelph, Ontario, Department of Biology, University of Ottawa, Ottawa, Ontario and Department of Zoology, University of Guelph, Guelph, Ontario (Canada)

INTRODUCTION

An essential prerequisite for the management of a renewable resource is a comprehensive understanding of its capabilities within the environmental confines conducive to growth and development. Present understanding of the growth of fish owes much to the conceptual models of Ivlev [1], Winberg [2], Paloheimo and Dickie [3,4], and Ursin [5]. These have served not only to demonstrate deficiencies but also as an impetus for many of the recent quantitative measurements made on the components of the energy schemes [6—12; and others].

This paper presents a review of the bioenergetic processes in fishes and environmental constraints on growth.

THE BIOENERGETIC SCHEME

An assessment of growth using a single criterion can sometimes be misleading. Growth studies which emphasize weight changes are concerned primarily with quantitative and preclude any qualitative changes. In contrast to this view is the use of the total energy concept. A more restricted assessment is the measurement of a single biological constituent such as nitrogen [7,13]. While each criterion has its merits, an integration of the different criteria would certainly provide a better understanding of growth. A change in biomass may involve simultaneous changes in proximate body composition and energetics while the converse may not necessarily be so. A bioenergetics scheme using the criterion of energy, but equally applicable for weight or protein, is presented in Fig. 1. The scheme is modified after those pro-

188

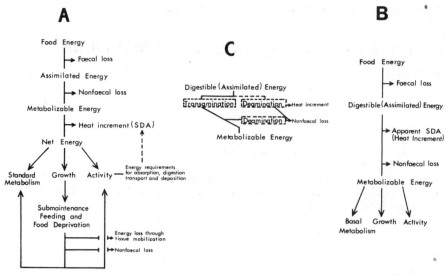

Fig. 1. Bioenergetics scheme of food utilization. Scheme A is similar to those generally accepted (after Niimi and Beamish [12]). Scheme B represents a revised version. The biochemical basis for the revision is also shown in Scheme C.

posed earlier by Ivlev [1], Winberg [2], Warren and Davis [14] and Brett [15]. The nomenclature is basically that employed by Harris [16] and Warren and Davis [14] except for the term non-faecal loss which for fish is a more appropriate term than urinary loss.

A reexamination of the scheme presented in Fig. 1 suggests it is not entirely compatible with the biochemical breakdown of food materials. A revised scheme that would be more consistent is described in Scheme B of Fig. 1, the difference being the position of the energy lost as apparent specific dynamic action [17] or heat increment. During digestion, the absorbed amino acid nutrients that are transported via the hepatic portal system must first be deaminated or transaminated, which occurs mainly in the liver. Following deamination the resultant compounds are available either for storage or utilization while the liberated NH groups may be excreted or retained for the synthesis of amino acids (Fig. 1C).

In both schemes, food energy is equivalent to the energy content or heat of combustion of the ingested food. The feeding level may range from deprivation to satiation or excess. Starvation has been used extensively to be synonymous with food deprivation. However, starvation is a clinical symptom [18] which may not be immediately apparent during the initial periods of food deprivation among fish. Satiation is the total amount of food ingested given all the food a fish would consume. Frequently satiation level is based not on a single daily meal but rather on two or more feeding intervals. Among largemouth bass, *Micropterus salmoides*, fed emerald shiners, *Notro-*

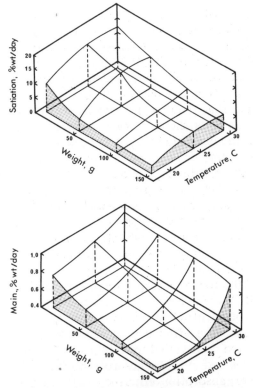

Fig. 2. Satiation and maintenance feeding levels of largemouth bass, *M. salmoides,* in relation to body weight and temperature [12].

pis atherinoides, satiation levels of feeding (% wet body wt/day) decreased curvilinearly with increasing body weights and decreasing temperature [12] (Fig. 2). A similar pattern with respect to satiation feeding levels and body weight was described for sockeye salmon, *Oncorhynchus nerka,* by Brett [10].

Faecal energy loss represents the energy content of undigested foodstuff. In practice it is difficult to separate faecal components from material sloughed from the lining of the gutwall and catabolised digestive enzymes. Maynard and Loosli [19] suggest the fraction of metabolic faecal protein in mammals, while difficult to measure, is usually not more than 0.1% dry wt of food consumed. When faeces is estimated by repetitive collection, a low estimate may arise from incomplete recovery. Where total collection of faeces is attempted it should be removed as soon as possible after defaeca-tion. For bass at 25°C, protein nitrogen content of faecal material allowed to remain in the tank water declined at a rate of about 18% over 24 h but changes in lipid were not detected [20]. A number of chemical indicators have been employed in digestibility studies to avoid the necessity for total collection of faecal material. The most widely used of such indicators is chromium sesquioxide. The success of chemical indicators depends in part on their homogeneity in the foodstuff as well as in the faeces and in part on their assay [21]. The rate at which ingested foodstuff is digested and the undigestible fraction evacuated appears relatively independent of quality [22] and quantity [20,23]. However, the time to evacuate the gut following ingestion of a large meal is greater than that required for small meals [20].

Assimilated or digestible energy represents the difference between food energy and that of faecal materials. In the bioenergetics scheme proposed by Warren and Davis [14], assimilated-energy is used rather than digestible energy. Preference for digestible energy in Scheme B is based on its earlier usage by domestic animal nutritionists [16] and its implication of being a less quantitative measure than is suggested by the term assimilated energy.

Digestible energy, expressed as a percentage of that ingested, is frequently designated as absorption efficiency, which in carnivorous fish varies from 85 to 98% [7,20,24—26,26a]. Observations among herbivores are limited but suggest a wider range of absorption efficiency. For example, Fischer [27] reported an absorption efficiency of 20% for grass carp, Ctenopharyngodon idella, and Menzel [28] a value of 80% for angelfish, Holacanthus bermudensis, fed vegetative material. Absorption efficiency among fish fed prepared diets also display wide variations. The efficiency of rainbow trout, Salmo gairdneri, fed 17 different prepared diets varied from 52 to 95% while for channel catfish, Ictalurus punctatus, fed 5 diets, absorption efficiency ranged from 12 to 80% [29]. However, absorption efficiency for commercially marketed feeds generally falls within the range of 80—90%.

Nonfaecal loss represents the excreted portion of digestible energy. This is comparable to the urinary loss of domestic animals and represents the energy content of the unutilizable portion of absorbed food materials. A major portion of this consists of the nitrogenous products of protein catabolism. In fishes, the term nonfaecal loss would be more precise since, in addition to the urinary loss of nitrogenous products such as creatine, uric acid, and some ammonia, significant amounts of diffusible products such as urea and ammonia are excreted across the gills. Smith [30] reported 6—10 times more nitrogen was excreted across the gills than by the kidneys in carp, Cyprinus carpio, and goldfish Carassius auratus. Catheterization of the bladder is the method most frequently employed in the collection of urine [31] while metabolic excretions across the epithelial surfaces can be determined from samples of appropriately treated bath water [26]. Where direct measurement is not made of nonfaecal loss, a good approximation can be estimated using the deductions of Krueger et al. [32]. Since the end-products of protein catabolism account for over 90% of excretory products [33] a nonfaecal loss of 15% of the energy derived from protein utilized would not be an unreasonable estimate when ammonia and urea are the primary end-products of protein catabolism [34].

Metabolizable energy represents the difference between food energy and the cumulative faecal and nonfaecal losses in the earlier scheme (A) but in the more recent scheme (B) specific dynamic action or heat increment is also deducted. Following ingestion of a meal, the rate of metabolism, expressed in units of heat production, increases. This increase is generally known as "specific dynamic action" but also as calorigenic effect [35], and heat increment [16]. The biochemistry of this process is not completely understood but liberated energy is generally assumed to be largely the result of the deamination of amino acids. The liver appears to be the primary if not the only site of heat production [36]. Although heat production is most pronounced with protein [37], some is released following ingestion of carbohydrate and fat. If each component is fed separately, their respective heat production in homeothermic animals would be equivalent to 30% of the caloric content of protein, 13% for lipid, and 5% for carbohydrate [38]. On the other hand, for a composite diet including protein, lipid, and carbohydrate, the total production of heat is less than that calculated from the nutritional composition of the feed [39].

The ingested proteinaceous material that is catabolized can be deaminated at either level of the bioenergetics scheme (Fig. 1), depending on the nutritional composition of the food and the physiological state of the animal. In view of this, it would not be appropriate to arbitrarily separate this portion of the bioenergetics scheme as was done previously (Fig. 1A). The term "metabolizable energy" as used in the revised scheme (Fig. 1B) is not empirically equivalent to that of the previous schemes, the difference being the position of the heat increment. However, it is probably the most accurate term in describing the energy remaining since it can either be anabolized or catabolized by the animal.

The energy liberated from heat increment is of little or no use to homeotherms except for maintaining body temperature in extreme cold [35]. Similarly, the poikilothermic response of fishes would suggest no substantial utilization of the energy liberated, except perhaps in some tunas and sharks which maintain elevated body temperatures [40]. This would further suggest heat increment should not be considered as a portion of metabolizable energy.

Energy requirements for absorption, digestion, transportation, and deposition of food materials while presumed to be minimal are distinct from those for specific dynamic action but experimentally difficult to separate. Where the distinction is not made the term apparent specific dynamic action is appropriate. Particularly little is known of apparent specific dynamic action in fish notable exceptions being the measurements made by Averett [41] on coho salmon *Oncorhynchus kisutch* and by Muir and Niimi [42] for aholehole, *Kuhlia sandvicensis*. Apparent specific dynamic action was determined by Beamish [17] for largemouth bass forced to swim at a fixed and low velocity in a tunnel respirometer [43]. The cumulative increase in O_2 consumption following ingestion of a prescribed meal was monitored. An estimate of the metabolic expenditure associated with the feeding procedure was made and substracted from the cumulative O_2 increase, the remainder representing apparent specific dynamic action (Fig. 3). Apparent specific dynamic action rose curvilinearly with ration size, the rate increasing with weight of bass. Thus over rations of 2—8% apparent specific dynamic action of a 25-g bass increased from 4.8—133.8 mg and for fish of 200 g from 218.7—989.5 mg. For a fixed feeding level, apparent specific dynamic action increased curvilinearly with size of bass, the rate being most pronounced at the higher levels of food intake. Conversion of apparent specific dynamic action from units of O_2 consumption to the standard heat equivalent is based either on complete oxidation of the foodstuffs or on the physiological equivalents of the metabolised substrates. Winberg [2] suggests that irrespective of the components oxidized the oxycalorific coefficient will not vary more than 1.5%. It is generally agreed an oxycalorific coefficient of 3.36—3.44 is most applicable for the higher vertebrates. For largemouth bass, apparent specific dynamic action expressed in energy units as a percent of ration intake did not differ significantly with weight of fish or meal size, the overall mean being 14.2% of the energy ingested.

Net energy of the widely accepted bioenergetics Scheme A, or the metabolizable energy of the revised Scheme B (Fig. 1), represents that portion of food energy that is available for other physiological activities. This energy can be distributed among three systems including basal or standard metabolism, growth, and locomotory activity. Basal metabolism and that associated

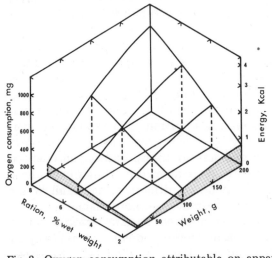

Fig. 3. Oxygen consumption attributable on apparent specific dynamic action of largemouth bass, *M. salmoides* in relation to meal size of lake emerald shiners, *Notropis atherinoides*, and body weight [17]. Conversion to the energy equivalent was made using the oxycalorific value of 3.42 cal/mg O_2.

with different rates of swimming is usually equated to O_2 consumption, the assumption being that all energy is released aerobically. Energy equivalents can be applied only if the system is totally aerobic. More precise estimates of metabolism particularly under the stress of high swimming speeds or burst performance can be obtained from simultaneous measurements of O_2 uptake and CO_2 production. Unfortunately few studies along these lines have been carried out with fish, the research of Kutty [44,45] on goldfish, rainbow trout and tilapia (*Tilapia mossambica*), being a prominent exception. The term standard metabolism was first employed by Krogh [46] as a measure approximating the basal rate. However, with the development of improved techniques in respirometry [43,47—51] measurements of O_2 consumption for fish in a post-absorbative condition and in a complete state of rest can now be made with precision. Where these techniques have been carefully applied it would seem appropriate to designate the metabolic level as basal rather than standard as suggested by Brett [52].

Swimming or locomotory activity can be divided into spontaneous and forced; the former usually termed routine [53] or intermediate [54]. Forced activity in the past usually approximated in short term or acute experiments is generally now measured over extended periods referred to as sustained swimming. The term active metabolism is defined as the maximum rate of O_2 consumption under continuous forced activity, again after a suitable period of food deprivation to assure the post-absorbative state and in the absence of external stimuli. O_2 consumption of fish forced to swim only over short intervals display much higher energy expenditures than those forced to swim over longer intervals at comparable velocities [55]. With training most fish perform in the swimming chamber with less effort and appear to be less excited [43,56,57]. Over extended periods physiological changes have been observed with training. Hochachka [58] found rainbow trout trained in a current for 6 months had higher haemoglobin levels and relatively larger hearts than untrained trout.

In most growth studies the assignment of an energy value for locomotory activity has been obtained by indirect means rather than by actual measurement. This can be largely overcome where growth experiments are conducted in annular tanks in which a known and fixed current of water is generated against which fish must continuously swim. Precaution must be taken to assign a swimming speed sufficiently high to prevent spontaneous activity but well below the active rate of O_2 consumption to allow for aerobic expansion of metabolism following ingestion of a meal.

Maintenance feeding is generally regarded as that level at which a change in weight will not occur [59]. Utilization of lipid as the principle energy source during deprivation has been established in the higher vertebrates. Lipid may contribute up to 85% of the energy requirements of some animals [60]. Nevertheless, some protein is still catabolized as a result of the metabolic processes [61]. The amount of protein catabolized will depend on the feeding level, proximate composition of the food and the protein sparing effect. If feeding is below the maintenance level, the food ingested will be catabolized. As feeding increases up to a maximum level, protein utilization will depend on the response of the proximate body composition of the fish on the feeding level. This response will vary among species. For example, protein content in largemouth bass remains relatively constant at 16.4% of body weight over an extended feeding range while protein levels in sockeye salmon appear to decrease from 18.6 to 11.4% of body weight as feeding levels are increased [62]. This difference, perhaps, should also be viewed as a response of an absolute deposition of biomass and its relative protein content.

Protein catabolism will also depend on the protein sparing effect, or the utilization of carbohydrate and fat to meet the energy requirements of an animal thereby making more protein available for proliferation of cellular components. This concept has been amply demonstrated in domestic animals [19], however, its significance has not been fully examined in fishes. The sparing effect of carbohydrate has been demonstrated in chinook salmon, *Oncorhynchus tshawytscha* [63] and channel catfish [64] fed prepared diets.

Protein utilization is also regulated by the nutritional composition of the food. Decreased protein digestibility have been correlated with increased carbohydrate content of the diet [65] and poor amino acid balance [66]. Hastings and Dickie [67] give a comprehensive account on the formulation and evaluation of prepared diets.

AN ASSESSMENT OF GROWTH USING THE BIOENERGETIC APPROACH, ITS PERSPECTIVES AND ITS PROBLEMS

The bioenergetics approach may profitably be viewed as the elaboration of the growth process, including negative growth, as measured by changes in biomass, proximate body composition, and energy content. This synthesis of several growth indices is essential towards a comprehensive examination of growth since it would include measurements on quantitative as well as qualitative changes in biomass. By applying the appropriate conversion factors, it would be possible to use indirect methods to derive estimates not readily attainable using conventional procedures. These factors, derived mainly from mammalian studies and almost universally applied, are the heat of combustion values of 9.4 and 5.65 kcal/g of fat (lipid) and protein, respectively, and the crude protein value of 6.25 multiplied by the nitrogen content [68]. The relationship between O_2 consumption and energy expenditure is generally accepted to be 3.36—3.44 cal/mg of O_2, the derivation of this value being based on a respiratory quotient of 0.8—0.9 (e.g. [11,14]).

A reexamination of these values as they relate to fishes suggest a revision would be in order before they can be effectively applied. Quantitative estimates of lipid is often determined by refluxing the sample with an organic solvent using a soxhlet or Goldfish apparatus. The quality of lipid extracted, may differ appreciably from the 9.4-kcal value as indicated by the 7.8 kcal/g (ash-free) found by Niimi [62] for largemouth bass. Even after purification of the extract using a phasing process (e.g. [69,70]), the heat of combustion was 8.5 kcal/g. There would be some loss of energy through autoxidation during the preparative and analytical processes, however, it would probably not account for the energy difference. It can further be suggested that the 9.4-kcal value is representative of the heat of combustion of the longer chained fatty acids and not the entire lipid complement. Presently there is insufficient evidence to support this view, however, a greater emphasis towards calorimetric measurements of the lipid extract may indicate a revised coefficient would be in order.

A similar observation can also be presented for protein. Crude protein is estimated using the 6.25 coefficient. This procedure would overestimate actual protein content since 9—18% of the total nitrogen is nonproteinaceous in those fishes examined [71,72]. Preliminary analysis of the fat free dried material of bass suggest the caloric content of the catabolizable portion, presumably consisting mainly of proteinaceous materials, to be 4.9 kcal/g (ash-free), and the nonproteinaceous portion considerably less [62]. Presently there is no suitable method for rapid and accurate evaluation of the caloric content of the protein and nonprotein nitrogenous fractions. A reliable quantitative estimate for protein would then require at least an adjustment for the nonproteinaceous fraction, and as experimental evidence becomes available, a reexamination of the energetics of the proteinaceous and nonproteinaceous constituents and a revision of the conversion factors as required.

The oxycalorific coefficient of 3.36—3.44 cal expended per mg of O_2 utilized is applicable during aerobic metabolism. When this coefficient is applied to the total catabolized energy, theoretically the intake energy minus faecal and nonfaecal losses, growth, and perhaps heat increment, an inbalance between the caloric expenditure and the oxygen requirement for carp, salmon, and bass is suggested at the higher feeding levels (Table I). The amount of O_2 required for total aerobic catabolism, based on a 3.42 value, in most cases approaches or exceeds the estimated maximum respiratory rates for the species examined. In contrast, at maintenance feeding, the calculated O_2 requirements for salmon and bass are only slightly above the basal metabolic level and well within aerobic limits. This would then strongly imply the relationship between the rate of O_2 consumption and the total energy catabolized may not be directly equated over the entire range of aerobic metabolism. An alternative energy catabolization scheme is an integration of the aerobic and anaerobic processes together with a change in metabolic efficiency [55].

TABLE I

RELATIONSHIP BETWEEN INGESTED ENERGY AND ITS UTILIZATION

An oxycalorific coefficient of 3.42 cal/mg consumed was applied in the calculations. Note the relation between O_2 required for aerobic catabolism and the active metabolic rate.

Species	Ivlev Mirror carp	Bret et al. Sockeye salmon		Niimi Largemouth bass			
Mean weight (g)	200	6.15	19.0	100	100	100	100
Temperature (°C)	20	15	15	25	25	30	30
Feeding level	Ad libitum	Maintenance	Excess	Maintenance	Excess	Maintenance	Excess
Diet	Chironomids	Prepared	Prepared	Emerald shiners		Emerald shiners	
Ingested energy (Kcal)	15.90	0.151	2.12	1.41	7.71	2.84	9.67
Faecal loss (Kcal)	3.95	0.032	0.42	0.23	1.26	0.49	1.49
Nonfaecal loss (Kcal)	4.98						
Growth (Kcal)	4.98	0.008	0.44	0.35	3.31	0.38	3.24
Energy catabolized (Kcal)	6.97	0.111	1.26	0.83	3.14	1.97	4.94
O_2 required for aerobic catabolism (mg/hr)	85	1.35	15.2	10.1	38.30	24.0	77.2
Standard (Basal) metabolism (mg/hr)	10*	0.98**	2.3***	8.0***	8.0***11.0***	11.0***	11.0***
Active metabolism (mg/h)		5.54**	17.1**	38.6**	38.60***		56.1***

* Extrapolated from Beamish and Dickie (55).
** Extrapolated from Brett (52).
***Extrapolated from Beamish (98).

With the experimental evidence presently available on fishes it is difficult to elaborate further. However, if the findings on vertebrate and invertebrate systems are taken into account, some alternatives to the probable biochemical mechanisms may be suggested. If the respiratory system of a fish cannot totally respond to its metabolic demands, the classical concept of O_2 debt is probable. The accumulation of lactic acid under these conditions is well established in higher vertebrates. This may be further pursued experimentally by examining the accumulation of lactate, propionate, acetate, and succinate, the end-products of anaerobic catabolism of carbohydrates, and the accumulation of alanine, proline, isobutyrate, isovalerate, and methylbutyrate, the end-products of anaerobic catabolism of amino acids observed in invertebrates [73]. Krueger et al. [32] suggested the probable excretion of these metabolites, however, Brett [11] does not support this view. It might not be appropriate at this time to reject one view in favor of the other since studies on the excretory products of fishes have been limited primarily towards the nitrogenous metabolites.

Another alternative is the dissipation of heat energy liberated in biochemical processes in accordance with the laws of thermodynamics. The quantity of energy attributable to this process can only be estimated but the loss cannot be avoided such energy loss would help to explain the increase in unaccountable energy as feeding levels increase.

A synthesis of both alternatives which would include a multiple end-product system and the accountability of the heat energy dissipated would perhaps be viewed most favorably. Of course, the acceptance or rejection of these mechanisms can only be examined through experimentation.

The recommendations suggested here are not totally consistent with the present views and procedures used in assessing the growth of fishes, although preliminary evidence does strongly suggest these views warrant further considerations. The flexibility, accountability, and accuracy of the bioenergetics approach will be recognized only after the reliability of the conversion factors can be verified through experimentation. If necessary, further modification of the bioenergetics scheme presented (Fig. 1) may be suggested as experimental evidence dictates.

ENVIRONMENTAL RESTRAINTS ON GROWTH

Growth is generally considered to reflect the net stress effects of environmental conditions on metabolic processes and as such may be viewed as a "wide spectrum indicator" [74]. Growth may respond not only to environmental influences on the processes of digestion, absorption, transportation, and deposition, but also through the amount of food consumed and on the level of locomotory activity. In the circumstance where an environmental identity acts to decrease the efficiency with which energy is accumulated, compensation may be achieved by increasing consumption of food. Thus where feeding level is not restricted or measured, the influence of the environment may be difficult to assess. Of course the same criteria apply for changes in the metabolic expenditure of energy due to alterations in activity induced by some environmental factor. Few investigative studies on the growth process of fish have controlled these factors sufficiently well to provide meaningful interpretation of the influence of the environmental factor under review.

Seasonal changes in caloric value of food items may be profound. Tyler [75] reported pronounced seasonal patterns in caloric content of a number of marine invertebrates which comprised the principal prey of a community of demersal fishes. For example, the caloric content of the krill, *Meganyctiphanes norvegica*, was at a minimum in April-May (4.4 kcal/g dry wt) and at a maximum in July (5.4 kcal/g) followed by a second minor peak in November-December.

Fish are not indiscriminate feeders, but rather display important differences in dietary components among species [76—81] and may undergo changes in food preferences with size [81]. Density of fish has been shown to influence growth. Overcrowding may lead to a reduction of food intake, growth rate, and to an increase in locomotory activity [82] while low densities may also impair growth [83—85]. The optimum density for growth may vary with age as Brown [83,84] demonstrated for brown trout, *Salmo trutta*. In contrast, Magnuson [86] found that medaka, *Oryzias latipes*, raised in different densities did not vary in growth rate provided other factors such as food and the elimination of metabolic products were not limiting. If food was in limited supply, growth depensation would occur.

Particle size of food items may also influence growth. Paloheimo and Dickie [3] proposed that brook trout, *Salvelinus fontinalis*, grew more rapidly when fed on large particles of food than small, the suggestion being that less energy was expended in searching for the large particles. Ivlev [87] reported that the consumption of food items by carp increased with the variance of distribution of food for a given feeding level.

Temperature is unquestionably one of the most influential environmental identities affecting the growth of fishes [8,12,88—95]. Nevertheless, few studies have examined growth of fish fed restricted rations in relation to temperature. No information appears available where both feeding level and activity were controlled.

Metabolic studies provide considerable insight into understanding the influence of temperature on growth. Basal metabolic requirements for fish of a given species and weight customarily increase with temperature along a convex curve with maximum rates related to thermal adaptation [96,97]. Thus for a constant food consumption, the proportion of energy required to meet basal needs increases with temperature. In accord, maintenance requirements of largemouth bass increased exponentially with temperature and body weight [12] paralleling changes in basal metabolic requirements. For sockeye salmon, Brett et al. [8] found maintenance requirements increased 7-fold between 1 and 20°C for fish of similar weight. The expenditure of energy for a given level of sustained swimming below maximum appears relatively independent of temperature [51,98], although at maximum sustained swimming (active metabolism), performance and the associated energy expenditure are temperature dependent. Subtraction of basal from active metabolism is termed "metabolic scope for activity" [53] and reflects the aerobic capacity of the species. Where measurements have been made, metabolic scope increases with temperature to an optimum and thereafter declines. Free swimming fish suggest marked changes in spontaneous activity in response to temperature. Spontaneous activity of the goldfish displayed a distinct response to the range of temperatures applied, 10—35°C, with the greatest value occurring at 20°C [50]. Thus, where activity is uncontrolled but meal size restricted, the energy expended in swimming might well be expected to vary with temperature leaving more or less energy available for growth.

In perhaps the most comprehensive study yet made on the influence of temperature on growth, Brett et al. [8] found optimum growth of small sockeye salmon occurred at approx. 15°C for the highest feeding level offered, declining progressively to a lower temperature with reduction in meal size. Gross conversion efficiency, the percent increase in growth relative to food consumption, followed a pattern similar to that for growth. Presumably the decrease in maintenance requirements that accompany a reduction in temperature permitted comparatively better growth at lower temperatures provided meal size was fixed and below satiation and the level of spontaneous activity remained relatively unchanged. This would apply only for poikilothermic animals and assumes the metabolic systems and digestive enzymes are reasonably eurythermal in their activity, a conclusion not entirely concordant with the observations on temperature optimum for the digestive enzymes of brown bullheads [99]. In contrast to observations on sockeye salmon, growth of largemouth bass [12] did not shift with temperature' when feeding level was altered but rather remained at a maximum at 18°C.

Winberg [2] suggested gross conversion efficiency was not influenced by temperature. In contrast, Brett et al. [8] indicated temperature may be as influential on gross conversion efficiency as feeding level and suggested the earlier conclusion appears to have resulted from a lack of observations at temperatures above the optimum. Where food is present in excess, temperatures below the optimum are associated with reduced food intake resulting in similar conversion efficiency.

Temperature influences not only growth but also the body caloric content and constituents. Among these, lipid and moisture appear to be most responsive to changes in temperature. Protein and ash levels do not change appreciably [12,100]. Decreasing temperatures enhance deposition of lipid

[8,12,101,102] an exception being channel catfish, *Ictalurus punctatus*, where lipid deposition increases at a near linear rate with increasing temperature [103]. The relations with moisture are less well understood. Iles and Wood [104] suggest in eviserated Atlantic herring, *Clupea harengus*, as lipid is depleted an equivalent weight of moisture is accumulated. A reexamination of estimates given by Brett et al. [8] and Groves [101] suggests that for sockeye salmon a decrease in lipid content is followed by a large increase in moisture content. On the other hand in largemouth bass, a decrease in lipid content is supported by a proportionately smaller increase in moisture content. This would probably account for the low weight loss of fishes subject to prolonged periods of food deprivation [62].

Dissolved O_2 is generally believed to restrict growth in low concentration [93,105—110] although few studies have been made where feeding level or activity were controlled. Some information on the role of dissolved O_2 on the growth process is apparent from metabolic investigations. The active metabolic rate of some species such as brook trout [11] and sockeye salmon [51] is dependent on dissolved O_2 to over 100% of air saturation. For others such as largemouth bass [98], carp, and goldfish [111] the active metabolic rate is independent of O_2 concentration, the range varying with the species. Basal metabolism is independent of dissolved O_2 over a wider range than that observed for active metabolism [112]. Among spontaneously active fish, it has been found that O_2 uptake for a given level of activity is less at low ambient O_2 than it is at air saturation [44,112,113]. However, the explanation that some fraction of the metabolism is liberated anaerobically in low ambient O_2 [112,113] was not supported by Kutty's measurements of respiratory quotients on goldfish and rainbow trout which remained at about one irrespective of ambient O_2. The biological mechanism whereby this is achieved is not known but of obvious ecological significance as it would extend the utilizable habitat and enhance growth in an O_2 poor environment.

In a comprehensive review on the dissolved oxygen requirements of fishes, Doudoroff and Shumway [54] suggest growth of post larval fish receiving an excess of food and maintained at suitable temperatures can be limited by low O_2 concentration. Further, growth may be depressed, although not as severely, in O_2 concentrations well above air saturation. Appreciable diurnal fluctuations in ambient O_2 may exert more influence on growth than continuous exposure to the mean O_2 concentration [109,114]. However, as suggested by Doudoroff and Shumway [54] where food is unrestricted, fluctuations in growth of fish in response to reduced O_2 may reflect changes in appetite or limitations imposed by the gas on exploitation of the food resource. Where fish are fed restricted meal sizes dissolved O_2 appears to exert little influence on growth. Thus growth of juvenile coho salmon of similar weight and fed restricted and equal rations showed no appreciable change in growth in concentrations of 3—18 mg O_2/l and was only slightly depressed at 3 mg O_2/l [114].

Photoperiod may influence growth although in temperate regions annual cycles of day length and temperature are nearly in phase and the individual role of these two factors are difficult to distinguish in the natural environment. In laboratory studies, food consumption and maintenance requirements of green sunfish, *Lepomis cyanellus*, increased with length of daylight [115,116]. Growth was generally highest in the increasing and lowest in the decreasing photoperiod although the influence of appetite and spontaneous activity was difficult to assess. Anderson [116] attributed the increase in maintenance requirements with length of photoperiod to an elevation in spontaneous activity. Further, Björklund [117] found goldfish grew better in darkness than in light which he attributed to a decrease in locomotory activity in the absence of light. No investigative studies on growth appear to have been made where activity and feeding level were measured nor on the influence of light intensity or quality. Gross et al. [115] suggested the changes in gross conversion efficiency which followed those of growth in relation to photoperiod reflected differences in protein synthesis brought about by photoperiod control of growth hormone production via the pituitary.

The role of salinity on growth has been examined in association with temperature by Kinne [82] and Saunders and Henderson [118]. Kinne [82] observed growth of desert pupfish, *Cyprinodon macularius*, to be intimately related to both salinity and temperature. Pupfish acclimated to 15 and 20°C grew fastest on unrestricted rations in % while for fish at 25—35°C, optimum growth was recorded in 35 and 55‰. Saunders and Henderson [118] exposed Atlantic salmon, *Salmo salar*, smolts and post-smolts to various salinities and temperature and monitored growth. During spring and early summer, growth was faster in 30 and 15‰ than in 0‰. Food consumption and gross conversion efficiency were higher in 30‰ then in 15 and 0% during spring and summer but fell rapidly during autumn and winter. In contrast to observations on the desert pupfish, temperature for optimum growth of Atlantic salmon was independent of salinity.

Salinity as a factor influencing the growth process of euryhaline species would appear to operate mainly through the expenditure of energy for osmoregulation. Conceivably, temperature, photoperiod, or any of a number of factors might act to modify the optimum salinity for osmoregulation. The energy expenditure for swimming in relation to salinity has been determined for rainbow trout [119] and *Tilapia nilotica* [43]. The actual expenditure for swimming in both species was independent of salinity, although, sizeable variation occurred in the cost of osmoregulation which was assumed to be zero at the isosmotic salinity. For *T. nilotica*, approx. 29% of the total O_2 uptake of swimming fish was required for osmoregulation at 30‰, and 19% at 0, 7.5, and 22.5‰ where isosmotic salinity was 11.6‰. Thus under conditions of restricted food supply and similar activity and without the influence of other environmental identities, growth might be expected to be maximal at the isosmotic salinity, decreasing progressively at both higher and lower concentrations.

Man's intervention with nature has recently stimulated research on the role of the disturbed environment on such sublethal responses as growth. The influence of various toxic materials in sublethal concentrations is perhaps best represented by the growth process which may be viewed as an expression of the net stress effects on metabolic processes. Exposure of sockeye salmon to sublethal concentrations of sodium pentachlorophenate demonstrated a reduction both in growth rate and gross conversion efficiency when fed restricted rations [74]. Pentachlorophenate is known to increase mitochondrial metabolism and to suppress glycolytic activity [120,121]. Elevated mitochondrial activity following exposure to pentachlorophenate resulted in high fat utilization in coho salmon [122] and in *Cichlasoma bimaculatum* [123]. Standard O_2 consumption of the guppy, *Lebistes reticulatus*, was increased in the presence of pentachlorophenate [124].

Similar patterns of growth change have been observed for sockeye salmon exposed to sublethal levels of kraft mill effluent [125] and for rainbow trout fed sublethal doses of endrin [126]. Observations on the endrin treated fish indicate an increase in swimming which could account for the reduction in growth. Recently P.F.K. Lett (unpublished data) measured growth of rainbow trout exposed to sublethal concentrations of reactive copper and fed restricted and equal rations of a prepared diet. Fish were forced to swim at a low swimming speed designed to negate differences in growth attributable to variations in activity. After an initial reduction in growth, the rate gradually increased so that after seven weeks exposure to concentrations as high as 50% of the 96-h LC_{50} it coincided with that of the control fish.

CONCLUDING REMARKS

Information on the bioenergetics and feeding behavior of fish has application in the interpretation and manipulation of biological productivity [127]. Until recently, the relationships between food intake and growth have been described only by inadequately small models which have little appreciation for the intermediate steps necessary to define growth [2,3,128]. Kerr [129] and Mackinnon [130] have attempted to predict production by analysing part of the bioenergetic scheme. However, little attention has been given to the feeding behavior of fish or the dynamics of the ecological community in which they belong.

Certainly one region of deficiency is in our knowledge of the methods by which fish capture prey, and the motivation which leads to searching. Some insight into the variables which control this behaviour have been established by Beukema [131], Magnuson [132], Brett [133], Ware [134], and Colgan [135]. However, our understanding of the relationship between search speed or the energy cost of searching and hunger is still fragmentary.

Emphasis on fish energetics has all too often been at the expense of examining the energetics of their food supply. The method by which energy is transferred among trophic levels is, of course, essential to the understanding of production.

With the information now available on the various components of the energy scheme together with improved programming techniques, computer models may contribute significantly to our general understanding. Perhaps the main benefit gained in model building is not so much in its predictive value but in the elucidation of areas in need of research.

REFERENCES

1. Ivlev, V.S. (1939) [Balance of energy in carps.] Zool. Zh. 18, 303–318. (In Russian with English summary.)
2. Winberg, G.G. (1956) The rate of metabolism and food requirements of fishes. Belorussian University, Minsk (In Russian, Fish. Res. Board Can. Transl. Ser. 194).
3. Paloheimo, J.E. and Dickie, L.M. (1965) Food and growth in fishes. I. A growth curve derived from experimental data. J. Fish. Res. Board Canada. 22, 521–542.
4. Paloheimo, J.E. and Dickie, L.M. (1966) Food and growth in fishes. III. Relations among food, body size, and growth efficiency. J. Fish. Res. Board Canada. 23, 1209–1248.
5. Ursin, E. (1967) A mathematical model of some aspects of fish growth, respiration and mortality. J. Fish. Res. Board Can. 24, 2355–2453.
6. Pandian, T.J. (1967) Intake, digestion, absorption and conversion of food in the fishes *Megalops cyprinoides* and *Ophiocephalus striatus*. Marine Biol. 1, 16–32.
7. Birkett, L. (1969) The nitrogen balance in plaice, sole and perch. J. Exp. Biol. 50, 375–386.
8. Brett, J.R., Shelbourn, J.E. and Shoop, C.T. (1969) Growth rate and body composition of fingerling sockeye salmon, *Oncorhynchus nerka*, in relation to temperature and ration size. J. Fish. Res. Board Can. 26, 2363–2394.
9. Gecking, S.D. (1971) Influence of rate of feeding and body weight on protein metabolism of bluegill sunfish. Physiol. Zool. 44, 9–19.
10. Brett, J.R. (1971) Statiation time, appetite, and maximum food intake of sockeye salmon (*Oncorhynchus nerka*). J. Fish. Board Can. 28, 409–415.
11. Brett, J.R. (1973) Energy expenditure of sockeye salmon, *Oncorhynchus nerka*, during sustained performance, J. Fish. Res. Board Can. 30, 1799–1809.
12. Niimi, A.J. and Beamish, F.W.H. (1974) Bioenergetics and growth of largemouth bass (*Micropterus salmoides*) in relation to body weight and temperature. Can. J. Zool. 52, 447–456.
13. Gerking, S.D. (1952) The protein metabolism of sunfishes of different ages. Physiol. Zool. 25, 358–372.
14. Warren, C.E. and Davis, G.E. (1967) Laboratory studies on the feeding, bioenergetics and growth of fish. In The Biological Basis of Freshwater Fish Production, (Gerking, S.D., ed.), pp. 175–214, Blackwell, Oxford and Edinburgh.
15. Brett, J.R. (1970) Fish — the energy cost of living. In Marine Aquiculture. (McNeil, W.J., ed.), pp. 37–52, Oregon State University Press, Oregon.

204

16. Harris, L.E. (1966) Biological energy interrelationships and glossary of energy terms. Natl. Acad. Sc., Washington.

17. Beamish, F.W.H. Apparent specific dynamic action of largemouth bass, Micropterus salmoides. J. Fish. Res. Board Can. 52(4), 447—456.

18. Blood, D.C. and Henderson, J.A. (1963) Veterinary Medicine. Bailliere, Tindall and Cox, London.

19. Maynard, L.L. and Loosli, J.K. (1969) Animal Nutrition. MacGraw-Hill, New York.

20. Beamish, F.W.H. (1972) Ration size and digestion in largemouth bass, Micropterus salmoides Lacepede. Can. J. Zool. 50, 153—163.

21. Czarnocki, J., Sibbald, J.R. and Evans, E.V. (1961) The determination of chromic oxide in samples of feed and excrete by acid digestion and spectrophotometry. Can. J. Animal Sci. 41, 167—179.

22. Windell, J.T. (1967) Rates of digestion in fishes. In The Biological Basis of Freshwater Fish Production, (Gerking, S.D., ed.), pp. 151—173, Blackwell, Oxford and Edinburgh.

23. Tyler, A.V. (1970) Rate of gastric empyting in young cod. J. Fish. Res. Board Can. 27, 1177—1189.

24. Gerking, S.D. (1955) Influence of rate of feeding on body composition and protein metabolism of bluegill sunfish. Physiol. Zool. 28, 267—282.

25. Menzel, D.W. (1960) Utilization of food by the Bermuda reef fish, Epinephelus guttatus. J. Cons. 25, 216—222.

26. Iwata, K. (1970) Relationship between food and growth in young crucian carps, Carassius auratus cuvieri, as determined by nitrogen balance. Jap. J. Limnol. 31, 129—151.

26a. Edwards, R.R.C., Finlayson, D.M. and Steele, J.H. (1972) An experimental study on the oxygen consumption, growth, and metabolism of the cod (Gadus morhua L.). J. exp. Marine Biol. Ecol. 8, 299—309.

27. Fischer, Z. (1970) The elements of energy balance in grass carp (Ctenopharyngodon idella Val.) - Part I. Polsk. Arch. Hydrobiol. 17, 421—434.

28. Menzel, D.W. (1959) Utilization of algae for growth by the Angelfish, Holacanthus bermudensis. J. Cons. 24, 308—313.

29. Halver, J.E. (1972) Fish Nutrition. Academic press, New York and London.

30. Smith, H.W. (1929) The excretion of ammonia and urea by the gills of fish. J. Biol. Chem. 81, 727—742.

31. Hunn, J.B. and Willford, W.A. (1970) The effect of anesthetization and urinary bladder catheterization on renal function of rainbow trout. Comp. Biochem. Physiol. 33, 805—812.

32. Kreuger, H.M., Sanddler, J.B., Chapman, G.A., Tinsley, I.J. and Lowry, R.R. (1968) Bioenergetics, exercise, and fatty acids of fish. Am. Zool. 8, 199—129.

33. Hoar, W.S. (1966) General and Comparative Physiology. Prentice-Hall, New Jersey.

34. Forster, R.P. and Goldstein, L. (1969) Formation of excretory products. In Fish Physiology, Vol. I. (Hoar, W.S. and Randall, D.J., eds), pp. 313—350, Academic Press, New York and London.

35. Kleiber, M. (1961) The Fire of Life, an Introduction to Animal Energetics. John Wiley and Sons, New York and London.

36. Buttery, P.J. and Annison, E.F. (1973) Considerations of the efficiency of amino acid and protein metabolism in animals. In The Biological Efficiency of Protein Production, (Jones, J.G.W., ed.), pp. 141—171, Cambridge University Press, London.

37. Krebs, H.A. (1964) The metabolic fate of amino acids. In Mammalian Protein Metabolism, Vol. I, Munro, H.N. and Allison, J.B., pp. 125—176. Academic Press, New York and London.

38. Harper, H.A. (1971) Review of Physiological Chemistry. Lange Medical Publications, Low Altos.

39. Forbes, E.B. and Swift. R.W. (1944) Associative dynamic effect of protein, carbohydrate and fat. J. Nutr. 27, 453—468.
40. Carey, F.G., Teal, J.M., Kanwisher, J.W., Lawson, K.V. and Beckett, J.S. (1971) Warm-bodied fish. Am. Zool. 11, 135—143.
41. Averett, R.C. (1969) Influence of temperature on energy and material utilization by juvenile coho salmon. Ph.D. thesis. Oregon State University, Corvallis, Oregon.
42. Muir, B.S. and Niimi, A.J. (1972) Oxygen consumption of the euryhaline fish, aholehole (*Kuhlia sandvicensis*), with reference to salinity, swimming, and food consumption. J. Fish. Res. Board Can. 29, 67—77.
43. Farmer, G.J. and Beamish, F.W.H. (1969) Oxygen consumption of *Tilapia nilotica* in relation to swimming speed and salinity. J. Fish. Res. Board Can. 26, 2807—2821.
44. Kutty, M.N. (1968) Influence of ambient oxygen on the swimming performance of goldfish and rainbow trout. Can. J. Zool. 46, 647—653.
45. Kutty, M.N. (1972) Respiratory quotient and ammonia excretion in *Tilapia mossambica*. Marine Biol. 16, 126—133.
46. Krogh, A. (1941) The quantitative relation between temperature and standard metabolism in animals. Int. Z. Phys. Chem. Biol. 1, 491—508.
47. Spoor, W.A. (1946) A quantitative study of the relationship between activity and oxygen consumption of the goldfish, and its application to the measurement of respiratory metabolism in fishes. Biol. Bull. 91, 312—325.
48. Ruhland, M.K. and Heusner, A. (1959) Chambre respiratorie pour la détermination simultane de l'activité et la consommation d'oxygène par une méthode manométrique chez des Poissons de 3—10 g. Soc. Biol. Paris, C. R. Séances 153, 161—164.
49. Blažka, P., Wolf, M. and Cepela, M. (1960) A new type of respirometer for the determination of the metabolism of fish in the active state. *Physiologia Bohemoslovaca* 9, 553—559.
50. Beamish, F.W.H. and Mookherjii, P.S. (1964) Respiration of fishes with special emphasis on standard oxygen consumption. I. Influence of weight and temperature on respiration of goldfish, *Carassius auratus*. Can. J. Zool. 42, 161—175.
51. Brett, J.R. (1964) The respiratory metabolism and swimming performance of young sockeye salmon, J. Fish. Res. Board Can. 21, 1183—1226.
52. Brett, J.R. (1972) The metabolic demand for oxygen in fish, particularly salmonids, and a comparison with other vertebrates. Resp. Physiol. 14, 151—170.
53. Fry, F.E.J. (1947) The effects of the environment on animal activity. University of Toronto Studies, Biological Series, No. 55. Publications of the Ontario Fisheries Research Laboratory, No. 68.
54. Doudoroff, P. and Shumway, D.L. (1970) Dissolved oxygen requirements of freshwater fishes. European Inland Fisheries Advisory Commission, Food and Aquicultural Organization of the United Nations. Technical Paper, 68.
55. Beamish, F.W.H. and Dickie, L.M. (1967) Metabolism and biological production in fish. In The Biological Basis of Freshwater Fish Production, (Gerking, S.D., ed.), pp. 215—242, Blackwell, Oxford and Edinburgh.
56. Brett, J.R., Hollands, M. and Alderdice, D.R. (1958) The effect of temperature on the cruising speed of young sockeye and coho salmon. J. Fish. Res. Board Can. 15, 587—605.
57. MacLeod, J.C. (1967) A new apparatus for measuring maximum swimming speeds of small fish. J. Fish. Res. Board Can. 24, 1241—1252.
58. Hochachka, P.W. (1961) Liver glycogen reserves of interacting resident and introduced trout populations. J. Fish. Res. Can. 18, 125—135.
59. Brown, M.E. (1957) Experimental studies on growth. In The Physiology of Fishes, (Brown, M.E., ed.), Vol. I, pp. 361—400, Academic Press, New York and London.
60. West, E.S. and Todd, W.R. (1964) Textbook of Biochemistry. MacMillan, New York.

61. Niimi, A.J. (1972) Changes in the proximate body composition of large-mouth bass (*Micropterus salmoides*) with starvation. Can. J. Zool. 50, 815—819.

62. Niimi, A.J. (1972) Bioenergetics and growth of largemouth bass (*Micropterus salmoides*) with feeding in relation to body weight and temperature. Ph.D. thesis. University of Guelph, Guelph, Ontario.

63. Buhler, D.R. and Halver, J.E. (1961) Nutrition of salmonid fishes. IX. Carbohydrate requirement of chinook salmon. J. Nutr. 74, 307—318.

64. Tiemeier, O.W., Deyoe, C.W. and Wearden, S. (1965) Effects of growth of fingerling channel catfish of diets containing two energy and two protein levels. Trans. Kansas Acad. Sci. 68, 180—186.

65. Kitamikato, M., Morishita, T. and Tachino, S. (1965) Digestion of dietary protein in rainbow trout. II. Effect of starch and oil contents in diets and size of fish on digestibility. Chem. Abstr. 62, 15129e.

66. Nose, T. (1963) Determination of nutritive value of food protein on fish. II. Effect of amino acid composition of high protein diets on growth and protein utilization of the rainbow trout. Bull. Freshwater Fish. Res. Lab. 13, 41—50.

67. Hastings, W.H. and Dickie, L.M. (1972) Feed formulation and evaluation. In Fish Nutrition (Halver, J.E., ed.), pp. 327—374, Academic Press, New York and London.

68. Brody, S. (1945) Bioenergetics and Growth, with Special Reference to the Efficiency Complex in Domestic Animals. Reinhold, New York.

69. Folch, J., Lee, M. and Stanley, G.H.S. (1957) A simple method for the isolation and purification of total lipids from animal tissues. J. Biol. Chem. 226, 497—509.

70. Bligh, E.G. and Dyer, W.J. (1959) A rapid method of total lipid extraction and purification. Can. J. Biochem. Physiol. 37, 911—917.

71 Simidu, W. (1961) Nonprotein nitrogenous compounds. In Fish as Food, (Borgström, G., ed.), pp. 353—384, Academic Press, New York and London.

72. Niimi, A.J. (1972) Total nitrogen, nonprotein nitrogen, and protein content in largemouth bass (*Micropterus salmoides*) with reference to quantitative protein estimates. Can. J. Zool. 50, 1607—1610.

73. Hochachka, P.W., Fields, J. and Mustafa, T. (1973) Animal life without oxygen: basic biochemical mechanisms. Am. Zool. 13, 543—555.

74. Webb, P.W. and Brett, J.R. (1973) Effects of sublethal concentrations of sodium pentachlorophenate on growth rate, food conversion efficiency, and swimming performance in underyearling sockeye salmon (*Oncorhynchus nerka*). J. Fish. Res. Board Can. 30, 499—507.

75. Tyler, A.V. (1973) Caloric values of some North Atlantic invertebrates. Marine Biol. 19, 258—261.

76. Todd, R.A. (1914) The food of the plaice. Fishery Investigations (Ministry of Agriculture and Fisheries). Series II. Salmon and Freshwater Fisheries. Vol. 2.

77. Richards, S.W. (1963) The demersal fish population of Long Island Sound. III. Food of the juveniles from a sand-shell locality. Bull. Bingham Oceanograph. Coll. Yale Univ. 18, 32—72.

78. Keast, A. (1965) Resource subdivision amongst coinhabiting fish species in a bay, Lake Opinicon. Great Lakes Research Division, University of Michigan, Publication 13, 106—132.

79. Nilsson, N.A. (1967) Interactive segregation between fish species. In The Biological Basis of Freshwater Fish Production (Gerking, S.D., ed.), pp. 295—313. Blackwell, Oxford and Edinburgh.

80. Zaret, T.H. and Rand, A.S. (1971) Competition in tropical stream fishes: support for the competitive exclusion principle. Ecology, 52, 336—342.

81. Tyler, A.V. (1972) Food resource division among northern, marine, demersal fishes. J. Fish. Res. Board Can. 29, 997—1003.

82. Kinne, O. (1960) Growth, food intake, and food conversion in a euryhaline fish exposed to different temperature and salinities. Physiol. Zool. 33, 288—317.

83. Brown, M.E. (1946) The growth of brown trout (*Salmo trutta* Linn.). II. The growth of two-year-old trout at a constant temperature of 11.5°C. J. Exp. Biol. 22, 130—144.

84. Brown, M.E. (1946) Growth of brown trout (*Salmo trutta* Linn.). III. Effect of temperature on the growth of two-year-old trout. J. Exp. Biol. 22, 145—155.

85. Paine, J.R. (1970) The influence of particle size and distribution of food on growth of the brown bullhead, *Ictalurus nebulosus* (Lesueur). M. Sc. thesis. University of Guelph, Guelph, Ontario.

86. Magnuson, J.J. (1962) An analysis of agressive behaviour, growth, and competition for food and space in medaka *Oryzias latipes* (Pisces, Cyprinodontidae). Can. J. Zool. 40, 313—363.

87. Ivlev, V.S. (1961) Experimental Ecology of the Feeding of Fishes. Yale University Press, New Haven.

88. Markus, H.C. (1932) The extent to which temperature changes influence food consumption in largemouth bass (*Huro floridana*). Trans. Am. Fish. Soc. 62, 202—210.

89. Haskell, D.C., Wolf, L.E. and Bouchard, L. (1956) The effect of temperature on the growth of brook trout. N.Y. Fish Game J. 3, 108—113.

90. Baldwin, N.S. (1957) Food consumption and growth of brook trout at different temperatures. Trans. Am. Fish. Soc. 86, 323—328.

91. Kramer, R.H. and Smith, L.L. (1960) First year growth of largemouth bass, *Micropterus salmoides* (Lacépede), and some related ecological factors. Trans. Am. Fish. Soc. 89, 222—233.

92. Strawn, K. (1961) Growth of largemouth bass fry and various temperatures. Trans. Am. Fish. Soc. 90, 334—335.

93. Swift, D.R. (1964) The effect of temperature and oxygen on the growth rate of the Windermere char (*Salvelinus alpinus willughbii*). Comp. Biochem. Physiol. 12, 179—183.

94. West, B.W. (1966) Growth rates at various temperatures of the orange-throat darter *Etheostoma spectabilis*. Arkansas Acad. Sci. Proc. 20, 50—53.

95. Lee, R.A. (1969) Bioenergetics of feeding and growth of largemouth bass in aquaria and ponds. M. Sc. thesis. Oregon State University, Corvallis, Oregon.

96. Scholander, P.F., Flagg, W., Walters, V. and Irving, L. (1953) Climatic adaptation in arctic and tropical poikilotherms. Physiol. Zool. 26, 67—92.

97. Fry, F.E.J. (1967) Responses of vertebrate poikilotherms to temperature. In Thermobiology (Rose, A.H., ed.), pp. 375—409, Academic Press, New York and London.

98. Beamish, F.W.H. (1970) Oxygen consumption of largemouth bass, *Micropterus salmoides*, in relation to swimming speed and temperature. Can. J. Zool. 48, 1221—1228.

99. Smith, H. (1967) Influence of temperature on the rate of gastric juice secretion in the brown bullhead (*Ictalurus nebulosus*). Comp. Biochem. Physiol. 21, 125—132.

100. Shul'man, G.Ye. and Kokoz, L.M. (1971) The content of fat-free dry matter in the body of some Black Sea fishes. J. Ichthyol. 2, 268—272.

101. Groves, T.D.D. (1970) Body composition changes during growth in young sockeye salmon (*Oncorhynchus nerka*) in freshwater. J. Fish. Res. Board Can. 27, 929—942.

102. Love, R.M. (1970) The Chemical Biology of Fishes. Academic Press. New York and London.

103. Andrews, J.W. and Stickney, R.R. (1972) Interactions of feeding rates and environmental temperature on growth, food conversion, and body composition of channel catfish. Trans. Am. Fish. Soc. 101, 94—99.

208

104. Illes, T.D. and Wood, R.J. (1965) The fat/water relationship in North Sea Herring (*Clupea harengus*), and its possible significance. J. Marine Biol. Assoc. U.K. 45, 353—366.
105. Nabiev, A.I. (1953) [Growth of young sturgeon, *Acipenser guldenstadti* Borodin, under experimental conditions and the influence on it of external environmental factors.] Trudy Inst. Zool. Akad. Nauk Az. SSR, Baku, 16, 87—147. (In Russian.)
106. Lozinov, A.B. (1956) [The oxygen optimum of sturgeon young.] Dokl. Akad. Navuk BSSR. 107, 337—339. (In Russian.)
107. Herrmann, R.B., Warren, C.E. and Doudoroff, P. (1962) Influence of oxygen concentration on growth of juvenile coho salmon. Trans. Am. Fish. Soc. 91, 155—167.
108. Switft, D.R. (1963) Influence of oxygen concentration on growth of brown trout, *Salmo trutta* L. Trans. Am. Fish. Soc. 92, 300—301.
109. Stewart, N.E., Shumway, D.L. and Doudoroff, P. (1967) Influence of oxygen concentration on the grwoth of juvenile largemouth bass. J. Fish. Res. Board Can. 24, 475—494.
110. Whitworth, W.R. (1968) Effects of diurnal fluctuations of dissolved oxygen on the growth of brook trout. J. Fish. Board Can. 25, 579—584.
111. Basu, S.P. (1959) Active respiration of fish in relation to ambient concentrations of oxygen and carbon dioxide. J. Fish. Res. Board Can. 16, 175—212.
112. Beamish, F.W.H. (1964) Respiration of fishes with special emphasis on standard oxygen consumption. III. Influence of oxygen. Can. J. of Zool. 42, 355—366.
113. Saunders, R.L. (1963) Respiration of the Atlantic cod. J. Fish. Res. Board Can. 20, 373—386.
114. Fisher, R.J. (1963) Influence of oxygen consumption and its diurnal fluctuations on the growth of juvenile coho salmon. M. Sc. thesis. Oregon State University, Corvallis, Oregon.
115. Gross, W.L. Roeloffs, E.W. and Fromm, P.O. (1965) Inlfuence of photoperiod on growth of green sunfish, *Lepomis cyanellus*. J. Fish. Res. Board Can. 22, 1379—1386.
116. Anderson, R.O. (1959) The influence of season and temperature on growth of the bluegill, *Lepomis macrochirus* Rafinesque. Ph.D. thesis. University of Michigan, Ann Arbor, Michigan.
117. Björklund, R.G. (1958) The biological function of the thyroid and the effect of length of day on growth and maturation of goldfish, *Carassius auratus* Linn, Ph.D. thesis, University of Michigan, Ann Arbor, Michigan.
118. Saunders, R.L. and Henderson, E.B. (1969) Growth of Atlantic salmon smolts and post smolts in relation to salinity, temperature and diet. Fish. Res. Board Can., Tech. Rep. 149, Western Pharmacol. Soc. 11, 129—132.
119. Rao, G.M.M. (1968) Oxygen consumption of rainbow trout (*Salmo gairdneri*) in relation to activity and salinity. Can. J. Zool. 46, 781—785.
120. Krueger, H. and Liu, S.D. (1970) Pentachlorophenol and acetate metabolism in *Cichlasoma bimaculatum*. Pharmacologist. 12, 208 pp.
121. Boström, S.L. and Johansson, R.G. (1972) Effects of pentachlorophenol on enzymes involved in energy metabolism in the liver of the eel. Comp. Biochem. Physiol. 41B, 359—369.
122. Hanes, D., Krueger, H., Tinsley, I. and Bond, C. (1968) Influence of pentachlorophenol on fatty acids of coho salmon (*Oncorhynchus kitsutch*). Proc. Western Parmacol. Soc. 11, 121—125.
123. Pasley, J.N., Chadwich, G.G. and Krueger, H. (1968) Thyroxine antagonism of pentachlorophenol poisoning in cichlid fish. Proc.
124. Crandall, C.A. and Goodnight, C.J. (1962) Effects of sublethal concentrations of

several toxicants on growth of the common guppy. Limnol. Oceanography. 7, 233—239.

125. Webb, P.W. and Brett, J.R. (1972) The effects of sublethal concentrations of whole bleached kraftmill effluent on the growth and food conversion efficiency of under-yearling sockeye salmon (*Oncorhynchus nerka*). J. Fish. Res. Board Can. 29, 1555—1563.

126. Grant, B.F. and Mehrle, P.M. (1970) Chronic endrin poisoning in goldfish, *Carassius auratus*. J. Fish. Res. Board of Can. 27, 2225—2232.

127. Dickie, L.M. (1973) Food chains and fish production. ICNAF. Special Publication 8, 201—221.

128. Ursin. E. (1967) A mathematical model of some aspects of fish growth, respiration, and mortality. J. Fish. Res. Board Can. 24, 2355—2453.

129. Kerr, S.R. (1971) Prediction of fish growth efficiency in nature. J. Fish. Res. Board Can. 28, 809—814.

130. Mackinnon, J.C. (1973) Analysis of energy flow and production in an unexploited marine flatfish population. J. Fish. Res. Board Can. 30, 1717—1728.

131. Beukema, J.J. (1968) Predation by the three-spined stickleback (*Gasterosteus aculeatus* L.). The influence of hunger and experience. Behaviour 31, 1—126.

132. Magnuson, J.J. (1969) Digestion and food consumption by skipjack tuna (Katsu-wonus pelamis). Trans. Amer. Fish. Soc. 98, 379—392.

133. Brett, J.R. (1971) Energetic responses of salmon to temperature. A study of some thermal relations in the physiology and freshwater ecology of sockeye (*Oncorhynchus nerka*). Am. Zool. 11, 99—113.

134. Ware, D.M. (1972) Predation by rainbow trout (*Salmo gairdneri*): the influence of hunger, prey density, and prey size. J. Fish. Res. Board Can. 29, 1193—1202.

135. Colgan, P. (1973) Motivational analysis of fish feeding. Behaviour. 45, 38—66.

136. Fry, F.E.J. (1957) The aquatic respiration of fish. In The Physiology of Fishes (Brown, M.E., ed.), Vol. I, pp. 1—63, Academic Press, New York and London.

Comparative Physiology — Functional Aspects of Structural Materials
Eds L. Bolis, H.P. Maddrell and K. Schmidt-Nielsen
© North-Holland Publishing Company — 1975 — Amsterdam

Effects of temperature and pressure on the oxidative metabolism of fish muscle

MALCOLM S. GORDON

Department of Biology and Institute of Evolutionary and Environmental Biology, University of California, Los Angeles, Calif. 90024 (U.S.A.)

SUMMARY

1. Maximal rates of O_2 consumption in vitro have been measured under standardized conditions on minced preparations of white muscle tissue from two groups of fishes: (a) at three temperatures (5, 15 and 25°C) on muscle from four species of mesopelagic fishes from the temperate Pacific Ocean; and (b) at the same three temperatures at each of four hydrostatic pressures (1400, 700 and 1000 atm) on muscle from both a temperate Pacific epipelagic fish and a tropical (though cool water) Pacific mesopelagic fish.

2. The four mesopelagic fishes, all members of the deep scattering layer community, had nearly identical, thermostable muscle O_2 consumption rates over the temperature range 5—15°C. They showed different Q_{10}'s over the range 15—25°C. The Q_{10} differences correlate fairly well with the probable ranges and magnitudes of thermal variations encountered during the vertical migrations of the species.

3. Muscle from both the cool water epipelagic fish and the cool water mesopelagic fish showed little metabolic sensitivity to high hydrostatic pressures at given temperatures. The rate versus temperature curve for the mesopelagic fish at 400 atm was peaked at 15°C.

4. Physiological and ecological implications of these results are discussed.

INTRODUCTION

There are many complexities associated with trying to find answers to the question, "Do deep-sea marine fishes show special metabolic adaptations for life under high hydrostatic pressures?" Two previous papers have discussed major aspects of these complexities. They also presented the results of series of measurements of O_2 consumption rates of substrate-saturated, minced preparations of white and red muscle tissues taken from an array of shallow-water marine fish species. These measurements established a base-line of data on shallow-water species against which data from deep-sea species can be compared [1,2].

An important conclusion reached in the first of these two papers was that, in order to isolate and identify specific metabolic responses to high pressures, comparisons between deep-sea and shallow-sea species of fishes must eliminate, as far as possible, the effects of other variables. More specifically, the most readily available deep-sea fishes, the vertically migratory, adult, mesopelagic fishes (such as lantern-fishes and other scattering layer forms) from areas with cool shallow water temperatures, would best be compared with cool water, adult, epipelagic, somewhat active or fairly active, shallow-sea species.

The present paper does exactly this, at least in part. Its major features are that it presents results from deep-sea, as well as shallow-sea fishes, and also presents data from series of experiments involving high hydrostatic pressures.

Experimental results of two kinds are presented: (1) Measurements of maximal O_2 consumption rates at three temperatures (5, 15 and 25°C) and normal (1 atm) hydrostatic pressure of minced preparations of white muscle (effectively mitochondrial preparations) from four species of mesopelagic deep-sea fishes from the temperate Pacific Ocean. Three species of lantern-fishes (Family *Myctophidae*) and one species of hatchet-fish (Family *Sternoptychidae*) are represented. (2) Measurements of maximal O_2 consumption rates at the same three temperatures at each of four hydrostatic pressures (1400, 700 and 1000 atm) of similar white muscle preparations from adults of a temperate Pacific species of active epipelagic fish (*Trachurus symmetricus*) and adults of a tropical (though cool water) Pacific species of mesopelagic deep-sea scorpionfish (*Ectreposebastes imus*).

The results permit discussion of aspects of the metabolic adaptations to both temperature and pressure shown by the white muscles of mesopelagic deep-sea as compared with a variety of shallow-sea fishes. The ranges of the two variables studied cover most of the oceanic ranges for these parameters. The results therefore provide information both of possible direct ecological relevance to the particular species studied and also reference data against which information we hope to obtain in future from more truly deep-sea species may be compared.

MATERIALS AND METHODS

White Muscle of Mesopelagic Fishes at Normal Pressure, Variable Temperature

Adult specimens of four species of mesopelagic deep-sea fishes were captured in the calm deep waters of the northeast lee of Guadalupe Island, México (near 29°N latitude; 118°W longitude) during February and March 1970. The four species represented two families (identifications confirmed by Robert Wisner, Scripps Institution of Oceanography). Body weight ranges for each species are given in parentheses: Family *Sternoptychidae: Argyropelecus pacificus* (2.0—6.8 g); Family *Myctophidae: Myctophum nitidulum*, (1.5—4.5 g), *Ceratoscopelus townsendi* (0.5—2.5 g) and *Triphoturus mexicanus* (0.1—4.9 g).

The *Myctophum* were captured at the sea surface at night by hand dip-net in the lighted area surrounding a 250 W water-proofed quartz iodide lamp (Hydro Products Model LQ-5) suspended just beneath the water surface. Fish captured were in excellent condition, and generally lost only a few scales due to handling.

The other three species were taken from hauls of a Tucker-type midwater trawl having an opening of 4 m^2. Hauls were made at ship speeds near two knots, at various times of day and night, with the most successful occurring during the early evening and early morning. Capture depths were uncertain, but were in the range 100—300 m. Total duration of individual hauls was usually 50—60 min, based upon 30-min hauling at maximum depth. Fishes captured were more or less battered in condition, having lost most of their scales and, frequently, a good part of their skins. Only specimens still significantly visibly physically active and in the best overall condition were used for experiments.

Hydrographic data at the Scripps Institution of Oceanography demonstrate the existence of environmental conditions, as given in Table I, in the uppermost 300 m of the sea in the area of our work during late February and early March.

Trawled fishes were kept in aerated surface sea water until sacrificed. They had white muscle sampels removed within 10 min after they arrived on board. Sampling, tissue preparation, and oxygen consumption measurement procedures were all as described by Gordon [1].

TABLE I

HYDROGRAPHIC DATA IN THE UPPERMOST 300 M OF THE SEA

Depth (m)	Temperature (°C)	Salinity (‰)	O$_2$ (ml/l)
0	15.5 — 18.0	33.5 — 34.0	5.5 — 5.7
50	15.5 — 17.5	33.5 — 34.0	5.3 — 5.6
100	13.0 — 17.5	33.6 — 33.9	5.2 — 5.4
200	8.5 — 10.1	33.8 — 34.2	2.0 — 3.5
300	7.5 — 9.2	34.1 — 34.2	0.9 — 2.2

White Muscle of Mesopelagic and Epipelagic Fishes at Variable Pressures and Temperatures

Adult specimens of the temperate Pacific, active epipelagic jack mackerel (*T. symmetricus*) in perfect condition were obtained from laboratory maintained stocks [3]. Fish weighed 100—300 g. The work was done during the months of May and June, when the fish were living at water temperatures near 18° C.

Adult specimens of the tropical Pacific, though cool water, mesopelagic deep-sea scorpionfish *E. imus* [4] were captured by night-time hauls of a 2-m width Isaacs-Kidd midwater trawl at depths of 150—300 m in James Bay, near Santiago (James) Island, Galápagos Islands, Ecuador (0° latitude, 91° W longitude), during October and November 1970. Trawling procedures were the same as were used at Guadalupe Island. Fish weighed 10—30 g and frequently came on board the research vessel in near perfect condition. They were maintained for up to 10 days in aquaria with aerated surface sea water cooled to 10° C. Only active, alert, nearly undamaged specimens were used for experiments.

Hydrographic conditions in the James Bay area at the time of this work are summarized by Sibert [5]. The waters at depths between 60 and 300 m had temperatures of 12—15° C and salinities of 34.8—35.0‰.

Tissue sampling and preparation procedures used were again as described by [1]. Maximal tissue O_2 consumption rates were measured on these samples in the high pressure respirometers described by Gordon and Thomas [6]. The main features of these operations were:

Weighed aliquots (500—1500 mg) of finely minced tissue were placed in each of two 100-ml capacity stainless steel pressure vessels filled with air equilibrated standard Ringer solution [1]. The sterilized vessels and the Ringer solution were previously equilibrated to the desired temperature for the particular run by near complete immersion in a thermoregulated refrigerated water bath fitted with a special mechanical magnetic stirring device. The pressure vessels were then closed, internal magnetic stirring started, and the system purged of air bubbles. Hydrostatic pressure in each vessel was set at desired levels, and tissue free samples of the enclosed Ringer (approx. 1.0—1.5-ml volume) were taken, without significant changes in pressure, at 10-min intervals over 1-h periods. The Ringer samples were rapidly transferred, without contact with the air, and flushed into and through the 75-μl volume sample chamber of a thermoregulated Instrumentation Laboratory Model 127 ultramicro oxygen electrode apparatus. This apparatus was connected to, and dissolved O_2 concentrations in the samples were determined with, an Instrumentation Laboratory Model 125-A portable polarographic analyzer. Precision of individual O_2 determinations was ±0.05 ml/l. O_2 consumption rates for tissue samples were calculated from the slopes of the straight lines relating dissolved O_2 concentrations to time. Duplicate runs under identical temperature and pressure conditions agreed within 5—10%.

Statistical Methods

Unless noted otherwise, statistical statements in this paper are based upon analysis of variance calculations. Significant differences are those for which $0.01 < P \leqslant 0.05$; highly significant differences are those for which $P \leqslant 0.01$.

RESULTS

White Muscle of Mesopelagic Fishes at Normal Pressure, Variable Temperature

Rate-temperature curves at normal (1 atm) pressure for substrate saturated white muscle preparations from the four species of mesopelagic fishes from Guadalupe Island are presented; in Fig. 1. Statistically highly significant variations in O_2 consumption rates occurred in three species, but no such variations occurred in the fourth, *T. mexicanus*. In the three species showing significant variations, these variations were restricted to the 15-25°C comparisons. Rates at 5 and 15°C were not significantly different from each other. For *M. nitidulum* the 15-25°C difference was significant; for the other two species the differences were highly significant. The Q_{10}'s for the 15-25°C interval for the three significantly varying species are: *A. pacificus*: 2.9; *M. nitidulum*: 1.4; *C. townsendi*: 2.6.

The three lanternfishes had identical O_2 consumption rates over the 5-15°C interval. The hatchetfish was significantly lower than the lanternfishes at 5°C, highly significantly lower at 15°C.

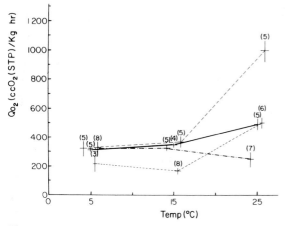

Fig. 1. Rates of O_2 consumption at three temperatures of white muscle preparations from four species of mesopelagic fishes from near Guadalupe Island, México. Species are: *M. nitidulum* (——); *C. townsendi* (— — —); *T. mexicanus* (— · — ·); *A. pacificus* (· · · · · ·). Horizontal lines at data points indicate means, vertical lines ±1 S.E., numerals the numbers of specimens in each data group.

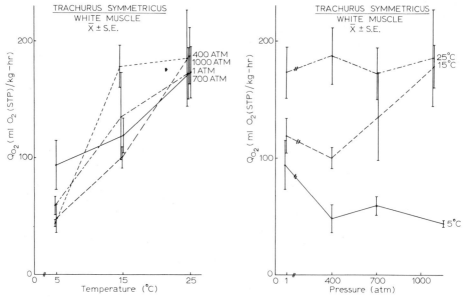

Fig. 2. Rates of O_2 consumption at three temperatures and four hydrostatic pressures of white muscle preparations from the southern California epipelagic jack mackerel, *T. symmetricus*. Data points are means for groups of 4—5 specimens, vertical lines ±1 S.E. In the lefthand graph, data at different pressures are indicated: 1 atm (————), 400 atm (— — —), 700 atm (— · — ·), 1000 atm (· · · · · ·). In the righthand graph, data at different temperatures are indicated: 5°C (————), 15°C (— — —), 25°C (— · — ·).

White Muscle of Epipelagic and Mesopelagic Fishes at Variable Pressure, Variable Temperature

Rate-temperature curves at four hydrostatic pressures for substrate-saturated white muscle preparations from the cool water epipelagic jack mackerel, *T. symmetricus*, are presented on the left side of Fig. 2. The right side of the figure replots the same data as rate-pressure curves for each of the three test temperatures.

Statistically significant variations in O_2 consumption rates occurred in the rate-temperature curves at 1 and 700 atm pressures, while highly significant variations occurred at 400 and 1000 atm. The variances of the different data groups are such that there is considerable inconsistency between the four pressure series in terms of their internal patterns of variation. This is illustrated by Table II.

However, excepting only the 1000-atm curve (flat between 15 and 25°C), all curves tend to be "normal" in shape (monotonically rising, with $Q_{10} >$ 1.3).

No statistically significant variations in O_2 consumption rates occurred in

TABLE II

RESULTS OF ANALYSIS OF VARIANCE OF TEMPERATURE DEPENDENCE OF O_2 CONSUMPTION RATES AT VARIOUS PRESSURES

Hydrostic pressure (atm)	Statistical comparison		
	5/15°C	15/25°C	5/25°C
1	N.S.	N.S.	*
400	*	*	**
700	N.S.	N.S.	**
1000	**	N.S.	*

N.S.: no statistical significance; *: significant difference; **: highly significant difference.

any of the three rate-pressure curves. However, at 15°C, the temperature nearest the normal environmental temperature for the experimental fish, pairwise comparisons between points demonstrate the existence of highly significant differences between the 1- and 400-atm points with respect to the 1000-atm point. Ratios for these rates are: 1000 atm/1 atm: 1.5; 1000 atm/400 atm: 1.8.

The indication thus is that the mitochondrial metabolism of the white muscle of the jack mackerel is insensitive to pressure changes at temperatures higher and lower than its adaptation temperature. It is similarly insensitive at moderately high pressures, but is activated by very high pressures at a temperature near its adaptation temperature.

Comparable sets of rate-temperature and rate-pressure curves for white muscle preparations from the cool water mesopelagic scorpionfish *E. imus* are presented in Fig. 3. Statistically significant variations in oxygen consumption rates occurred at 1 and 400 atm, but no significant variations (though the *F*-value is close to the $P = 0.05$ level) at 1000 atm. The 1 and 700 atm curves are statistically flat in the range 15—25°C. The 400-atm curve is peaked at 15°C (the test temperature nearest the normal environmental temperature). Elevated pressures thus appear to first increase, then inhibit the temperature sensitivity of these preparations.

No statistically significant variations in O_2 consumption rates occurred in the rate-pressure curves at 5°C and 15°C, while significant variations occurred in the data at 25°C. At 25°C the 400-atm point is slightly but significantly below both the 700- and 100-atm points. The overall pattern is one of insensitivity to pressure, but with a general trend at all three temperatures towards small magnitudes of metabolic activation at moderate and very high pressures.

DISCUSSION AND CONCLUSIONS

Previous experimental metabolic studies of deep-sea fishes have virtually

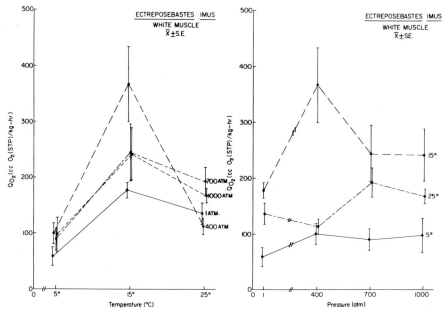

Fig. 3. As Fig. 2, but for white muscle preparations from the Galápagos mesopelagic scorpionfish *E. imus*.

all concentrated at the biochemical and cell physiological levels. Recent papers are from: Blaxter et al. [7], Childress [8], Gillen, [9]*, Hochachka et al. [10]*, Hochachka [11]*, Hochachka et al. [12]*, Meek and Childress [13]*, Nevenzel et al. [14], Nevenzel [15], Patton and Thomas [16], Phleger and Benson [17], Phleger [18]* and Smith and Hessler [19]*; (items marked with an asterisk contain results of experiments specifically involving elevated hydrostatic pressures). A recent review is by Flügel [20].

Aside from the data presented here, there appears to be nothing known about the effects of temperature upon the metabolism of the tissues of mesopelagic fishes, or about the combined effects of temperature and hydrostatic pressure upon the metabolism of the tissues of epipelagic and mesopelagic fishes. The present data provide insight into the nature of possible mitochondrial level adaptations in the muscles of these two groups of fishes to the ecologically significant ranges of these two variables. They also provide the beginnings of a base of data relating to more extreme conditions, similar to deep-sea conditions, with which future measurements on more truly abyssal forms may be compared. It should be remembered that the measurements reported were made using high, saturating substrate concentrations.

White Muscle of Mesopelagic Fishes at Normal Pressure, Variable Temperature

The four species of mesopelagic fishes studied all belong to the deep scattering layer community. Little is known about the behavior of the particular species involved. It is apparent, however, that the *Myctophum* is the species successfully coping with the widest range of environmental conditions, since it occurs in large numbers at the sea surface at night. Its exact daytime depth range is unknown, but it is unlikely to remain at depths shallower than 200 m.

The *Triphoturus* is probably the next most adaptable species. Ebeling et al. [21] indicate that its vertical distribution at night centers in the uppermost 300 m (our experience indicates the uppermost 200 m is probably more precise), while its daytime distribution is concentrated in the range 450—650 m. The *Ceratoscopelus* appears to stay at greater depths, its night distribution usually remaining below 200 m or so. Finally, indirect evidence cited by Ebeling et al. [21] makes it probable that the *Argyropelecus* does not migrate vertically to a significant extent, and lives primarily in the 250—350-m depth range.

The experimental results are reasonably consistent with these patterns of vertical distribution. The *Triphoturus* is the most consistently thermostable of the four, the *Myctophum* and *Argyropelecus* are next, and the *Ceratoscopelus* is least stable, especially in the 15—25° C range. The *Ceratoscopelus* probably never encounters temperatures much above 10° C, hence has not evolved effective metabolic control mechanisms for higher temperature ranges. The similarity of all species in the 5—15° C range fits in with the fact that all must cope with this range. Their uniform insensitivity to temperature change in this range presumably assists them to successfully carry out whatever vertical movements may be necessary without major changes in their efficiency or speed of swimming.

Comparison of the data for these four mesopelagic species with previous data for cool water shallow-sea species from southern California [1] shows that white muscle metabolic rates for the mesopelagic group are not significantly different, at any of the three study temperatures, from rates measured in adults of all shallow-sea species, as a group (11 species in the group). Comparisons with adults of the littoral, benthic, and epipelagic subgroups of the shallow-sea group, taken separately (3, 2, and 6 species in each group, respectively), also show no significant differences, except for the littoral/mesopelagic comparison at 25° C. At 25° C this difference is highly significant. The group mean metabolic rates are 280 and 500 ml/kg · h, respectively. The difference disappears, however, if the *Ceratoscopelus* data are removed from the mesopelagic group. The indications are, therefore, that the muscles of the relatively shallow-living groups of mesopelagic deep-sea fishes we studied are not significantly more cold-adapted than are cool water shallow-sea fishes.

White Muscle of Epipelagic and Mesopelagic Fishes at Variable Pressure, Variable Temperature

The most striking feature of the pressure-temperature results for the epipelagic *Trachurus* and the mesopelagic *Ectreposebastes* is that the pressure results are so similar. The jack mackerel was chosen for study on the basis of the statistical analysis of white muscle metabolism data for shallow-sea fishes described by Gordon [1]. Adults of this species, an active, cool water, epipelagic form, are among the group of shallow-sea fishes most likely to have the closest overall metabolic resemblance to adults of cool water, mesopelagic deep-sea species. Very little is known of the behavior and vertical depth distribution of *Ectreposebastes*, so this particular mesopelagic form may turn out to be unrepresentative of mesopelagic fishes in general. It is phylogenetically unique [4]. However, the fact that the two forms show nearly identical white muscle metabolic responses to high pressures raises the possibility that, at least at the mitochondrial level of structural complexity, the deep-sea group lacks any special pressure adaptations.

The rate-temperature curve for *Trachurus* at 1 atm is comparable in shape to the curve for the same species, obtained by a different respirometric method, included in Gordon [1]. The present curve, however, is much lower, in terms of absolute rates of O_2 consumption, than the previous curve. Reasons for this difference are not apparent. The only important differences between the two sets of measurements, in terms of tissue handling procedures, are that the present data derive from tissue samples 3—5 times heavier than those used in the earlier studies, and that the samples were dispersed into much larger volumes of air-equilibrated Ringer solution than were used earlier (100 ml compared with 4 ml).

The effects of elevated pressures upon temperature sensitivity of the *Trachurus* preparations are too small to be of any ecological significance to the species in terms of its normal shallow depth range.

The peaked shape of the rate-temperature curve for *Ectreposebastes* (at 400 atm) is duplicated in only one species of shallow-sea fish studied previously. This is the cool water, active, epipelagic top smelt, *Atherinops affinis* [1]. *Atherinops* is primarily a surface feeding fish, probably restricted to only the shallowest depths. The possible ecological value of a peaked white muscle rate-temperature curve to *Atherinops* is not apparent.

Ectreposebastes, however, may derive some ecological advantage from the unusual shape of its white muscle rate-temperature curve. This would especially be so if the curve was actually broadly, rather than sharply, peaked, and if the peaking became significant at pressures near 50 atm. Hydrographic conditions in the Galápagos are such that this fish never encounters temperatures higher than about $18°C$. If it does not migrate too extensively vertically, it may also rarely, if ever, encounter temperatures much below $10°C$. A relative insensitivity of muscle energy metabolism to temperature change within this range, especially an insensitivity combined with a maximal metabolic rate, could help ensure high swimming speeds and efficiencies throughout its daily vertical travels. The fact that intermediate levels of high pressure (400 atm), even though the specific test pressure levels were well above the ecologically significant range, increased the height of the 15°C peaking, may indicate some degree of reciprocal metabolic influence of increased pressure and lowered temperature in the deeper parts of the vertical range of the species. A reciprocal relationship would further serve to stabilize the swimming capabilities of the fish.

There is presently no way to determine the degree to which these white muscle results coincide with, or differ from, the metabolic properties of intact fishes under various combinations of temperature and pressure. The lack of whole animal data also precludes meaningful comparison with the information now available relating to vertically migratory epipelagic and mesopelagic crustacea [22].

ACKNOWLEDGEMENTS

This work has been supported by U.S. National Science Foundation research Grants GB-3584, GB-5661, GB-15180, and GB-31180; the Zoology-Fisheries program, University of California, Los Angeles (UCLA); and the R/V Alpha Helix Program, Scripps Institution of Oceanography, La Jolla, Calif. Field work was carried out with the cooperation and assistance of the Directors and staffs of Fishery-Oceanography Center, U.S. National Marine Fisheries Service, La Jolla, Calif.; and R/V Alpha Helix (Guadalupe Island cruise, 1970; Galapagos cruise, 1970). Computing assistance was obtained from the Health Science Computing Facility, UCLA, sponsored by NIH Grant FR-3. Technical assistance was provided by, in alphabetical order: B. Gordon, K. Tan, T. Thomas, J. Waggoner III, and D. Yang.

REFERENCES

1. Gordon, M.S. (1972) Comparative studies of the metabolism of shallow-water and deep-sea marine fishes. I. White muscle metabolism in shallow-water fishes. Marine Biol. 13, 222—237.
2. Gordon, M.S. (1972) Comparative studies of the metabolism of shallow-water and deep-sea marine fishes. II. Red muscle metabolism in shallow-water fishes. Marine Biol. 15, 246—250.
3. Hunter, J.R. and Zweifel, J.R. (1971) Swimming speed, tail beat frequency, tail beat amplitude and size in jack mackerel, *Trachurus symmetricus*, and other fishes. Fish. Bull. Natl. Marine Fish. Serv. 69, 253—266.
4. Eschmeyer, W.N. and Collette, B.B. (1966) The Scorpion-fish subfamily Setarchinae, including the genus *Ectreposebastes*. Bull. Mar. Sci. 16, 349—375.

5. Sibert, J. (1971) Some oceanographic observations in the Galápagos Islands. Am. Zool. 11, 405—408.
6. Gordon, M.S. and Thomas, T.J. (1974) Apparatus for studies of tissue preparations under high hydrostatic pressures. Marine Biol., 28, 73—77.
7. Blaxter, J.H.S., Wardle, C.S. and Roberts, B.L. (1971) Aspects of the circulatory physiology and muscle systems of deep-sea fishes. J. Mar. Biol. Assoc. U.K. 51, 991—1006.
8. Childress, J.J. (1971) Respiratory rate and depth of occurrence of mid-water animals. Limnol. Oceanogr. 16, 104—106.
9. Gillen, R.G. (1971) The effect of pressure on muscle lactate dehydrogenase activity of some deep-sea and shallow-water fishes. Marine Biol. 8, 7—11.
10. Hochachka, P.W., Schneider, D.E. and Kuznetsov, A.P. (1970) Interacting pressure and temperature effects on enzymes of marine poikilotherms: catalytic and regulatory properties of FDPase from deep and shallow-water fishes. Marine Biol. 7, 285—293.
11. Hochachka, P.W. (1971) Pressure effects on biochemical systems of abyssal fishes: the 1970 Alpha Helix expedition to the Galápagos Archipelago. Am. Zoologist 11, 399—576.
12. Hochachka, P.W., Moon, T.W. and Mustafa, T. (1972) The adaptation of enzymes to pressure in abyssal and midwater fishes. In: The effects of Pressure on Organisms (Sleigh, M.A. and Macdonald, A.G., eds), pp. 175—195, Academic Press, New York.
13. Meek, R. and Childress, J. (1973) The effect of hydrostatic pressure on the respiratory rate of a mesopelagic fish (*Anoplogaster cornuta*). Deep-sea Res. 20, 1111—1118.
14. Nevenzel, J.C., Rodegker, W., Robinson, J.S. and Kayama, M. (1969) The lipids of some lantern fishes (family *Myctophidae*). Comp. Biochem. Physiol. 31, 25—36.
15. Nevenzel, J.C. (1970) Occurrence, function, and biosynthesis of wax esters in marine organisms. Lipids 5, 308—319.
16. Patton, S. and Thomas, A. (1971) Composition of lipid foams from swim bladders of two deep ocean fish species. J. Lipid Res. 12, 331—355.
17. Phleger, C.F. and Benson, A.A. (1971) Cholesterol and hyperbaric oxygen in swimbladders of deep sea fishes. Nature 230, 122.
18. Phleger, C.F. (1972) Cholesterol and hyperbaric oxygen in swimbladders of deep sea fishes. Unpubl. Ph.D. dissertation, Scripps Inst. Oceanogr., La Jolla, Calif.
19. Smith, K.L. and Hessler, R.R. (1974) Respiration of benthopelagic fishes: in situ measurements at 1230 meters. Science 184, 72—73.
20. Flügel, H. (1972) Pressure: animals. In Marine Ecology (Kinne, O. ed.) pp. 1407—1437, Wiley-Interscience, London.
21. Ebeling, A.W., Ibara, R.M., Lavenberg, R.J. and Rohlf, F.J. (1970) Ecological groups of deep-sea animal off southern California. Bull. Los Angeles County Mus. Nat. Hist. Sci., 6, 1—43.
22. Teal, J.M. (1971) Pressure effects on the respiration of vertically migrating decapod crustacea. Am. Zool. 11, 571—576.

Comparative Physiology — Functional Aspects of Structural Materials
Eds L. Bolis, H.P. Maddrell and K. Schmidt-Nielsen
© North-Holland Publishing Company — 1975 — Amsterdam

Adrenergic control of blood flow through fish gills: Environmental implications

L. BOLIS and J.C. RANKIN

Institute of General Physiology, Laboratory of Comparative Physiology, University of Rome, Rome (Italy) and Department of Zoology, University College of North Wales, Bangor, North Wales (U.K.)

INTRODUCTION

When faced with an increased energy demand, as for example when swimming, fish increase O_2 uptake across the gills [1]. The flow of water over the external surfaces of the gills is increased as is the circulation of blood within the gill lamellae, which function as counter-current exchangers [2]. The circulatory response, involving an increase in cardiac output (all blood leaving the fish heart passes directly to the gills via the ventral aorta) and the redistribution of blood within the gills, can be used as a model for studying the effect of pollutants.

To stress the importance of the adrenergic control of blood flow, experiments were performed in vivo, and in vitro using perfused gills where the effect of a detergent (sodium alkylbenzenesulphonate) on the response to noradrenaline was studied in brown trout and eel.

MATERIAL AND METHODS

The experiments in vivo were performed in European eels *Anguilla anguilla* L. with chronically implanted cannulae in the dorsal and ventral aortae and in the pneumatic duct vein (for injection). The arterial cannulae were connected to pressure transducers (SE Laboratories, Type SEM-4-86 coupled to a transducer/converter Type SE 905) and the pressures were recorded on a potentiometric recorder (Telsec Instruments, Type 700T).

The in vitro experiments were performed on perfused gills with the technique previously described by Rankin and Maetz [3]. The effect on red blood cell permeability to D(+)-glucose of trout kept in presence of a detergent (sodium alkylbenzenesulphonate), was tested with Wilbrandt's direct and indirect methods [4,5]. Animals employed were either eels (*A. anguilla* L.) or trout (*Salmo trutta* L. or *S. gairdneri* Rich.) weighing 1 kg and 300—400 g, respectively; the substances employed were sodium alkylbenzenesulphonate (prevelance of C_{12}-alkyl chains), D(+)-glucose, adrenaline bitartrate, noradrenaline, phenylephrione, and isopunaline.

RESULTS AND DISCUSSION

In Vivo Studies on Fish Circulation

During moderate swimming activity the heart rate of rainbow trout increases only slightly, but the stroke volume shows a 5-fold rise [6]. Blood pressure in the ventral aorta, before the gills, and in the dorsal aorta, after the gills, is considerably elevated [1]. Randall and Stevens [7] found that the pressure increase during forced swimming in the salmon was similar to that produced by intravenous injection of adrenaline; in both cases the response could be prevented by treatment with the β-adrenergic blocking agent phenoxybenzamine. Similar responses to injected adrenaline and noradrenaline are seen in the eel [8,25] and cod [9]. In the rainbow trout, activity is associated with plasma catecholamine (mainly adrenaline) levels of up to 0.15 μg/ml compared with resting levels of less than 0.01 μg/ml [10]. The amount of catecholamine released into the blood stream during the first few min of swimming must have been of the order of several μg/kg body weight. In the injection studies referred to above the amounts injected were within and below this range.

The effects produced by such doses are dramatic and are sufficient to account for the circulatory effects observed during exercise. For example Fig. 1 shows the effect of injecting adrenaline (0.5 μg/kg body wt) into a fresh-water adapted eel. The upper trace shows the ventral aortic pressure, the upper and lower extremities of the trace representing the systolic and diastolic pressures and its width being proportional to the pulse pressure. The lower trace records dorsal aortic pressure. As the blood has passed through the gills the pressure has fallen and the amplitude of the pulse pressure has been attenuated. The extent to which this happens depends on the resistance of the branchial vasculature. The response to adrenaline (or noradrenaline, Fig. 2, which in this case shows the effects more clearly) injection into the pneumatic duct vein is immediate. Both pressures rise but the response is greater in the dorsal aorta; the pressure drop across the gills diminishes. The pulse pressure increases as the heart pumps more blood at each beat and the attenuation of the pulse pressure by the gills is less. This suggests that, since the blood flow through the gills has increased, the branchial vascular resistance has decreased considerably. On the other hand,

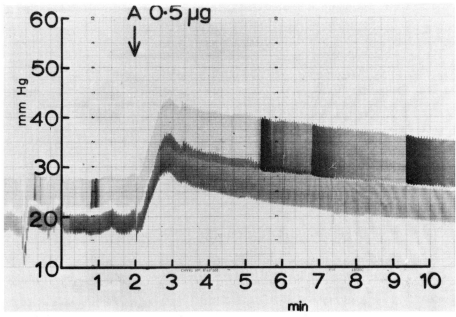

Fig. 1. Recording of ventral (upper trace) and dorsal (lower trace) aortic blood pressure in an eel before and after intravenously injection of 0.5 μg adrenaline (A).

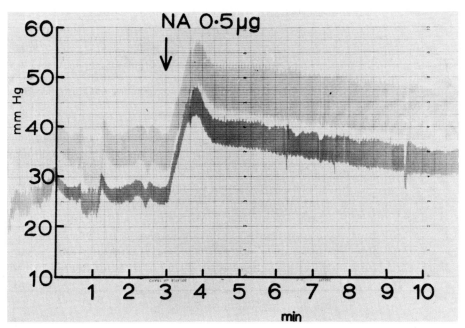

Fig. 2. Recording of ventral (upper trace) and dorsal (lower trace) aortic blood pressure in an eel before and after intravenously injection of 0.5 μg noradrenaline (NA).

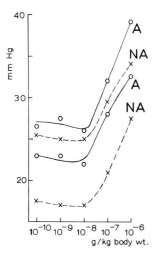

Fig. 3. Log dose response curve for pressor actions of adrenaline (solid lines) and noradrenaline (broken lines) in the same eel on different days. In each case the upper line represents mean ventral and the lower line mean dorsal aortic blood pressure. Note the reduction in the differential pressure produced by the 1 μg/kg dose of noradrenaline; in most cases this also applied for adrenaline but on the day the adrenaline curves were obtained the control pressure difference was lower than usual.

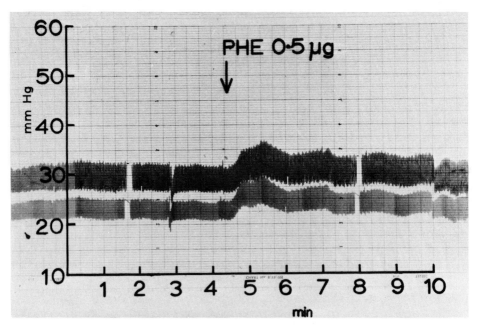

Fig. 4. Recording of ventral (upper trace) and dorsal (lower trace) aortic blood pressure in an eel before and after intravenously injection of 0.5 μg phenylephrine (PHE).

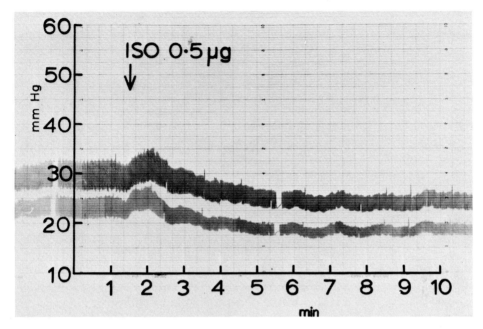

Fig. 5. Recording of ventral (upper trace) and dorsal (lower trace) aortic blood pressure in an eel before and after intravenously injection of 0.5 μg isoprenaline (ISO).

it has long been known that adrenaline produces systemic vasoconstriction in the eel [11] and this would account, in part, for the pressor effects. The potencies of adrenaline and noradrenaline are very similar (Fig. 3).

Receptors responding to catecholamines can be divided into two main groups; α- and β-receptors. The former usually cause vasoconstriction. The response to the drug phenylephrine which stimulates α- but not β-receptors is shown in Fig. 4. Dorsal and ventral aortic pressures rise to the same extent suggesting an increased vascular resistance distal to the gills. The pressure increases would be accentuated if the vasoconstriction was accompanied by increased blood flow. This is what happens with adrenaline and noradrenaline. The drug isoprenaline stimulates only β-receptors and its use might be expected to clarify the situation. However, after a transient increase it produces a prolonged decrease in both ventral and dorsal aortic pressures (Fig. 5) presumably by dilating systemic vessels [8,9]. During the experiments it is important to avoid any disturbance of the animal which can lead to vagal inhibition of the heart [26].

In Vitro Studies on Fish Gills

A number of studies on isolated perfused gill arches and perfused heads including the gills have shown vasodilation by catecholamines; some have included the use of α- and β-adrenergic agonists and antagonists to obtain information on the type of receptors present in the gills (for references see Wood [12]). In most of these studies dose-response curves were not obtained. One problem is the variation in sensitivity between different gills; there is also the difficulty of rinsing out the effects in order to obtain sufficient results for a dose response curve from one gill [3,13]. However, these problems have been overcome in a teleost, the rainbow trout [12] and an elasmobranch, the lesser spotted dogfish, *Scyliorhinus canicula* [14]. In the dogfish, noradrenaline, acting through α-receptors; isoprenaline, acting through β-receptors; and adrenaline, acting through both types, cause vasodilation. Wood also produced evidence for the presence of α-dilatory receptors in the trout gill, branchial vascular resistance being decreased by low concentrations of phenylephrine but he discounted this, as the maximum response was less than half that produced by adrenaline. He showed that β-receptors were present, the potencies of isoprenaline, noradrenaline and adrenaline being in the ratio 1000 : 33 : 10, calculated from log concentration response curves.

Plasma from stressed dogfish contains sufficient catecholamine to produce a marked dilation when added to the perfusion fluid in microliter amounts in the preparation of Davies and Rankin [14]. Concentrations of catecholamine similar to those found in resting eel blood produce a considerable increase in flow of Ringer solution through perfused eel gills [3] suggesting that tonic vasodilation in the gill vasculature may be maintained by circulating hormone levels. In the cod, *Gadus morhua*, stimulation of the branchial branches of the vagus nerve produces some increase in vascular resistance in a perfused gill [15]. A similar slight effect is seen in perfused elasmobranch gills (Davies, D.T., unpublished observations). A number of workers have found acetylcholine to be vasoconstrictory in both teleost (see [16]) and elasmobranch [14] gills. These effects are unlikely to be of great importance in vivo since they are slight, and there is no good evidence of innervation within the secondary lamellae of teleosts [17] where an important part of the regulation of blood flow may take place [16,18] (but see [2]).

The preparation of Rankin and Maetz [3] as modified by Davies and Rankin [14] cannot be used to calculate the actual vascular resistance of the gill as, although the perfusion is from a constant head apparatus, the pressure at the gill may vary as flow rate changes, the resistance of the fine bore polythene tubing used to reduce dead space to a minimum not being negligible in comparison with that of the gill. It does have the advantage of simplicity and, provided Millipore-filtered Ringer solution is used, the gills show no signs of increasing vascular resistance over many hours at room temperature. Although, in teleosts, recovery from catecholamine action is very slow compared to the dogfish, it has recently been possible to obtain cumulative dose response curves such as that shown for noradrenaline in the eel (Fig. 6). In this experiment each point on the graph is the mean of the values ob-

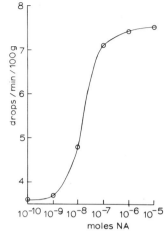

Fig. 6. Log dose-response curve for increase in flow through isolated perfused eel gill produced by noradrenaline (NA). Each point is the mean of the results from three different gills.

tained from three different gills. As long as the rate of flow was adjusted to the same value, relative to the body weight of the fish, before adding the hormone, consistent results were obtained. Flow rate could be adjusted by altering inlet or outlet pressures, which were kept close to normal physiological values. In eels, rainbow trout and brown trout this preparation gives different results to that of Wood [12] in that, although noradrenaline is slightly more potent than adrenaline, isoprenaline is less potent than either in spite of experiments with blocking agents which indicate that β-receptors predominate. It is possible to obtain repeated cumulative dose response curves from a single gill if adequate time is allowed for recovery from the effect of the highest dose of each hormone used. For example, in Fig. 7 is shown a dose response curve for noradrenaline obtained from a brown trout gill which was virtually identical to a second curve obtained after a 45-min interval.

The fact that gills treated in this manner will retain their sensitivity to catecholamines for several hours means that the preparation can be used to study the effects of environmental factors on gill adrenergic receptors. One topic of current interest is the effect of detergent pollution in water on fish. Detergents are toxic to fish at low concentrations, causing obvious histological damage to the gills, but little is known about possible deleterious effects of sublethal levels on physiological parameters, such as modification of blood flow through gills [19,20]. Since detergents may affect cell membrane structure, and β-adrenergic receptors are an integral component of the adenyl cyclase system incorporated in cell membranes [21], it seemed logical to look at the effect of detergents on the response of perfused gills to catecholamines.

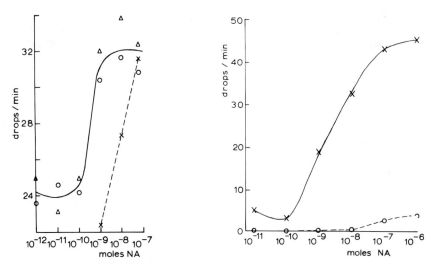

Fig. 7. Log dose-response curve for increase in flow through isolated perfused brown trout gill produced by noradrenaline (NA). The circles represent the first points obtained; the triangles the second set after an interval of 45 min. One curve fits both sets of points. After an interval of 32 min, at the beginning of which sodium alkylbenzenesulphonate (1 mg/l) was added to the Ringer solution, a third set of points (crosses) was obtained from which a second curve was drawn (broken line).

Fig. 8. Log dose-response curves for increase in flow through isolated perfused brown trout gill produced by noradrenaline (NA). After the control curve (solid line) had been obtained sodium alkylbenzenesulphonate (1 mg/l) was added and after 1 h the second curve (broken line) was obtained.

Sodium laurylsulpahte, a linear alkylate sulphonate, was added to the perfusion fluid at a concentration of 1 mg/l or 0.5 mg/l. The most obvious effect was a steady diminution in flow rate in both eel and brown trout gills. Fig. 7 shows the results of an experiment in which sodium alkylbenzenesulphonate was added to the Ringer solution after two similar dose response curves had been obtained for noradrenaline. The control rate of flow before each of these curves was the same, 21.8 drops/min. After 32 min perfusion with sodium alkylbenzenesulphonate it had fallen to 12.0 drops/min. At this point 10^{-10} mol of noradrenaline was added, increasing the flow rate to 19.3 drops/min and then the other points shown in Fig. 7 were obtained. The sensitivity had been reduced but the same maximum dilation was obtained. This was not the case after longer exposure to sodium alkylbenzenesulphonate. Fig. 8 shows an experiment in which, after a noradrenaline dose response curve had been obtained, sodium alkylbenzenesulphonate (1 mg/l) was added and the perfusion continued for 1 h. During this time the control rate of Ringer solution flow dropped from 3.6 to 0.5 drops/min. As can be seen from the dose-response curve, as well as reduced sensitivity, the maxi-

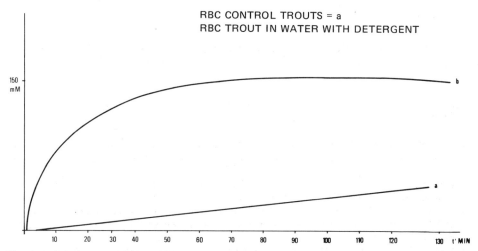

Fig. 9. D(+)-glucose transport in trout red blood cells in the presence of 3 mg/l sodium alkylbenzenesulphonate (24 h).

mum effect produced by very high doses of noradrenaline was very much reduced. Using a concentration of detergent of 0.5 mg/l the effect is less dramatic, showing that a dose relation can be derived. Higher doses of sodium alkylbenzenesulphonate dramatically inhibit any response to noradrenaline.

It might be argued that the detergent should have been added to the water in which the gills were suspended. There are two reasons why this procedure was not adopted. Firstly, unless vigorous external water circulation is maintained through the gill filaments, entry of substances from the external medium is much slower than in vivo; this may be because a film of mucus forms over the gill. Secondly, since the blood in fish gills is in constant close proximity to the water even slow entry of any pollutant could lead to significant accumulation in the blood. The results show that if detergent enters the blood stream (and experiments on red blood cell permeability of trout kept in water containing low levels of sodium alkylbenzenesulphonate indicate that it does) it will have serious adverse effects on gill adrenergic receptors.

Indeed, red blood cells from trout do not usually transport hexoses, such as D(+)-glucose [22—24], but in our experiments, following the direct and indirect method of Wilbrandt, using red blood cells from trout kept in water containing sodium laurylsulphate (3 mg/l), transport of D(+)-glucose was evident (Fig. 9). This effect is probably due to the fact that the action of the detergent can modify the structure of the red blood cell membrane, considering that the blood in the fish gills is in close contact with water.

Experiments on fish from polluted and non-polluted waters would be expected to show if this effect is produced by concentrations of detergent below those which cause obvious histological damage.

232

ACKNOWLEDGEMENT

Part of this work was supported with C.N.R. Grant No. 72.001101.11.115.
7030.

REFERENCES

1. Stevens, E.D. and Randall, D.J. (1967) Changes in blood pressure, heart rate, and breathing rate during moderate swimming activity in rainbow trout. J. Exp. Biol. 46, 307—315.
2. Hughes, G.M. and Morgan, M. (1973) The structure of fish gills in relation to their respiratory function. Biol. Rev. 48, 419—475.
3. Rankin, J.C. and Maetz, J. (1971) A perfused teleostean gill preparation: Vascular actions of neurohypophysial hormones and catecholamines. J. Endocrinol. 51, 621—635.
4. Wilbrandt, W. (1955) Osmotische Erscheinungen und osmotischen Methoden an Erythrozyten. Handbuch der Physiologische und Pathologische Chemische Analyse. 10. Aufl. Band II.
5. Wilbrandt, W., Fuhrmann, G.F. and Liggenstorfer, P. (1971) Änderung der Glukose Transport Frischer menschlicher Erythrozyten bei längere Inkubation. Experientia 27, 1428.
6. Stevens, E.D. and Randall, D.J. (1967) Changes of gas concentrations in blood and water during moderate swimming activity in rainbow trout. J. Exp. Biol. 46, 329—337.
7. Randall, D.J. and Stevens, E.D. (1967) The role of adrenergic receptors in cardiovascular changes associated with exercise in salmon. Comp. Biochem. Physiol. 21, 415—424.
8. Chan, D.K.O. (1967) Hormonal and haemodynamic factos in the control of water and electrolyte fluxes in the European eel, Anguilla anguilla L. Ph.D. thesis, University of Sheffield.
9. Helgason, S.S. and Nilsson, S. (1973) Drug effects on pre- and post-branchial blood pressure and heart rate in a free-swimming marine teleost, Gadus morhua. Acta Physiol. Scand. 88, 533—540.
10. Nakano, T. and Tomlinson, N. (1967) Catecholamine and carbohydrate concentrations in rainbow trout (Salmo gairdneri) in relation to physical disturbance. J. Fish. Res. Bd. Can. 24, 1701—1715.
11. Keys, A. and Bateman, J.B. (1932) Branchial responses to adrenaline and pitressin in the eel. Biol. Bull. 63, 327—336.
12. Wood, C.M. (1974) A critical examination of the physical and adrenergic factors affecting blood flow through the gills of the rainbow trout. J. Exp. Biol. 60, 241—265.
13. Randall, D.J., Baumgarten, D. and Malyusz, M. (1972) The relationship between gas and ion transfer across the gills of fishes. Comp. Biochem. Physiol. 41A, 629—637.
14. Davies, D.T. and Rankin, J.C. (1973) Adrenergic receptors and vascular responses to catecholamines of perfused dogfish gills. Comp. Gen. Pharmacol. 4, 139—147.
15. Nilsson, S. (1973) On the autonomic nervous control of organs in teleostean fishes. In Comparative Physiology: Locomotion, Respiration, Transport and Blood, (Bolis, L., Schmidt-Nielsen, K. and Maddrell, S.H.P., eds), pp. 323—331, North Holland, Amsterdam.
16. Richards, B.D. and Fromm, P.O. (1969) Patterns of blood flow through filaments and

lamellae of isolated-perfused rainbow trout (*Salmo gairdneri*) gills. Comp. Biochem. Physiol. 29, 1063—1070.

17. Morgan, M. and Tovell, P.W.A. (1973) The structure of the gill of the trout *Salmo gairdneri* (Richardson). Z. Zellforsch. 142, 147—162.

18. Steen, J.B. and Kruysse, E. (1964) The respiratory function of teleostean gills. Comp. Biochem. Physiol. 12, 127—142.

19. Abel, P.D. (1974) Toxicity of synthetic detergents to fish and aquatic invertebrates. J. Fish. Biol. 6, 279—298.

20. Marchetti, R. (1965) Critical review of the effects of synthetic detergents on aquatic life. General Fisheries Council of the Mediterranean Studies and Rev. No. 26, FAO, Rome.

21. Robison, A.G., Butcher, R.W. and Sutherland, E.W. (1971) Cyclic AMP. Academic Press, New York and London.

22. Bolis, L., Luly, P. and Baroncelli, V. (1971) D(+)-Glucose permeability in brown trout *Salmo trutta* L. erythrocytes. J. Fish Biol. 3, 273.

23. Bolis, L. and Luly, P. (1972) Membrane lipid pattern and non electrolyte permeability in *Salmo trutta* L. red blood cells. In Passive Permeability of Cell Membrane (Kreuzer, F., ed.), p. 357. Plenum Press, London.

24. Bolis, L. and Luly, P. (1972) Non-electrolyte permeability in *Salmo trutta* L. red blood cells. In Role of Membranes in Secretory Processes (Bolis, L., Katchalsky, A. and Keynes, R.D., eds), p. 215, North Holland, Amsterdam.

25. Maetz, J. and Rankin, J.C. (1969) Quelques aspects du rôle biologique des hormones neurohypophysaires chez les poissons. Colloq. Int. Cent. Nat. Rech. Sci. 177, 45—54.

26. Randall, D.J. (1968) Functional morphology of the heart in fishes. Am. Zool. 8, 179—189.

Comparative Physiology — Functional Aspects of Structural Materials
Eds L. Bolis, H.P. Maddrell and K. Schmidt-Nielsen
© North-Holland Publishing Company — 1975 — Amsterdam

Energy costs of ion pumping and the drain on the cell's metabolism

D.G. KILBURN, M.P. MORLEY and J. YENSEN

Department of Microbiology, University of British Columbia, Vancouver 8 (Canada)

I. INTRODUCTION

The energy utilized by mammalian cells for ion pumping has conventionally been estimated by measuring the amount of ATP hydrolysed, O_2 consumed or heat produced over a relatively short time interval, by very specialized cells (e.g. red cells, nerve fibres) and depends on the fact that these cells are not growing, since in a growing cell a sizeable proportion of the energy used would be directed toward biosynthesis rather than transport. The aim of this communication is to show how overall ion pumping energy requirements can be estimated from the rate of consumption of substrate and the cell yield in growing cultures.

II. THE CONCEPT OF MAINTENANCE ENERGY

It has been shown that in exponentially growing cultures of bacteria [1,2] and mammalian cells [3] the rate of energy generation from the catabolism of substrate can be expressed as the sum of a growth rate dependent component and a growth rate independent component.

$$\frac{dE}{dt} = A \frac{dx}{dt} + mx \tag{1}$$

where dE/dt = rate of ATP generation from substrate (which is assumed to be equal to its rate of utilization),

x = concentration of organism in terms of dry weight or cell number,

A, m = constants.

The growth rate dependent component accounts for the increase in cell mass, and represents the amount of energy used for biosynthesis, while the remaining energy, which depends only on the weight of cells present and not upon the rate at which they are growing, may be envisaged as serving house-keeping or maintenance functions. The foremost of these, in all likelihood, is the maintenance of concentration gradients by means of ion pumping. Since during exponential growth, $dx/dt = \mu x$, where μ = the specific growth rate, Eqn (1) may be integrated to give a relationship between the cumulative amount of energy generated and time during exponential growth.

$$E = (m + A\mu)\frac{x_0}{\mu}(e^{\mu t} - 1) \tag{2}$$

where x_0 = the concentration of cells at the onset of exponentional growth, t = time from the onset of exponential growth.

A plot of E vs $e^{\mu t} - 1$ should be a straight line passing through the origin with a slope equal to $(m + A\mu)x_0/\mu$. If this slope is determined for several different values of μ, the values of m and A can be calculated.

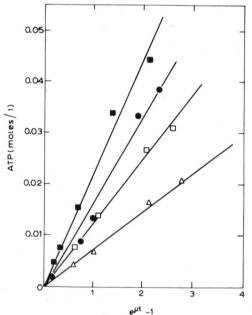

Fig. 1. Energy produced in batch cultures of LS cells grown at controlled dissolved O_2 partial pressures plotted against $e^{\mu t} - 1$: ■, pO_2 = 1.6 mm Hg, x_0 = 2.7 · 10^8 cells/l, μ = 0.30 day^{-1}; ●, pO_2 = 160 mm Hg, x_0 = 2.8 · 10^8 cells/l, μ = 0.53 day $^{-1}$; □, pO_2 = 96 mm Hg, x_0 = 2.4 · 10^8 cells/l, μ = 0.60 day^{-1}; △, pO_2 = 12 mm Hg, x_0 = 1.6 · 10^8 cells/l, μ = 0.65 day $^{-1}$. (Taken from Kilburn et al. [3].)

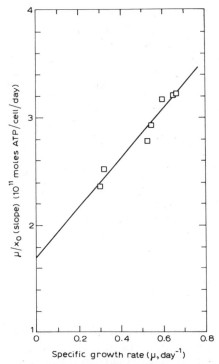

Fig. 2. Plot of specific growth rate, μ, against the term $\mu(\text{slope})x_0$ from Eqn (2). The value of the slope is taken from the lines in Fig. 1; x_0 is the cell concentration at the beginning of exponential growth. (Taken from Kilburn et al. [3].)

Fig. 1 shows a plot of this sort for the energy generated in batch cultures of mouse LS cells grown at different exponential growth rates by controlling the dissolved O_2 partial pressures in the medium. To calculate the energy generated, it was assumed that each mole of glucose oxidized completely gave 38 mol of ATP, while glucose metabolized to lactate gave 1 mol of ATP per mol of lactate produced. It can be seen that these data do indeed fall on straight lines passing through the origin. Rearranging the expression for the slope of these lines

$$\frac{\mu\,(\text{slope})}{x_0} = m + \mu A \tag{3}$$

gives an expression which allows m and A to be determined. A plot of the left hand side of this equation against μ is shown in Fig. 2. This is a straight line (with a slope equal to A) extrapolating to m when the growth rate is zero giving the following values:

$m = 1.7 \cdot 10^{-11}$ mol ATP/cell per day,

$A = 2.3 \cdot 10^{-11}$ mol ATP/cell.

From Eqn (1), the biosynthetic rate component is μA, which, if expended for the mean generation time ($t = (\ln 2)/\mu$), gives the energy to synthesize one new cell. i.e. $A \ln 2 = 1.6 \cdot 10^{-11}$ mol ATP/cell synthesized. Paul [4], knowing the composition of an L cell and assuming that all the monomers were available in the medium, has calculated from purely theoretical considerations that $1.15 \cdot 10^{-11}$ mol ATP would be required to synthesize one cell of the mass found in the above experiments. Hence the theoretical and experimental values are in good agreement. On the basis of our experimental figure, the maximum possible yield of cells per mol of ATP ($Y_{ATP,max}$) would be 42 g dry wt/mol ATP.

The maintenance energy coefficient of $1.7 \cdot 10^{-11}$ mol ATP/cell per day compares well with less reliable estimates of $1.9 \cdot 10^{-11}$ to $2.4 \cdot 10^{-11}$ mol ATP/cell per day derived from measurements on cells in stationary phase, which although non-growing, may have a substantial biosynthetic component from turnover which would increase the apparent value of m.

III. ENERGY FOR ION PUMPING

The energy expended on the regulation of ion concentration should be independent of growth rate, and thus comprise part of the maintenance energy as outlined above. Any change in the pumping load would thus be reflected by a change in the value of m. Therefore, to obtain a gross estimate of the effect of increasing, for instance, Na^+ concentration, a series of cultures could be grown in the presence of excess Na^+ at various growth rates regulated by the dissolved O_2 partial pressure and the graphical analysis of Figs 1 and 2 applied to determine the new value of m. The increase should relate to the overall effect of increasing the Na^+ pumping load. In fact, since the energy for biosynthesis can be considered to be constant, the slope of the line in Fig. 2 will not change, although its position and y intercept will shift upwards. Hence it is not necessary to perform a series of experiments at various controlled growth rates. It is sufficient to utilize one set of data obtained at the growth rate observed when pO_2 is not limiting, provided that this growth is exponential. This analysis would not hold of course, if toxic conditions were imposed, in which case growth would not usually be exponential.

Before such an analysis of the gross energetics of ion pumping can be contemplated, it is necessary to establish whether or not the value of m derived from batch cultures in which the physiological state of the organism is constantly changing (e.g. under conditions that vary from excess substrate to starvation) are truly representative. One means of doing this is to fix the growth rate and physiological state of the organism by growing it in continuous culture.

IV. THE USE OF CONTINUOUS CULTURE TO ESTIMATE m AND A

In continuous culture a suspension of growing cells is supplied with a steady flow of fresh medium and an equivalent volume of culture fluid (cells + depleted medium) is withdrawn to maintain a constant volume. Provided that the dilution rate, D, (defined as the flow rate of incoming medium divided by the volume of the vessel) does not exceed the maximum growth rate of the organism, the cells regulate their own growth rate to equal the dilution rate ie. $\mu = D$. [5]. Fig. 3 indicates how the rate of energy utilization varies with dilution rate in a continuous culture and shows the two previously defined components of energy expenditure. The maintenance component is seen to consume progressively more of the total energy available as the growth rate decreases, until at zero growth rate, the rate of energy generation equals m. Variations in m would be expected to shift this line up or down without affecting the slope. While this relationship is suitable for determining values of m for A for bacterial cultures, the range of dilution rates available for animal cell experiments is much narrower because of their slower growth rates, which limits the maximum dilution rate at which washout of the culture occurs and the minimum rate at which degeneration of medium components, and increased influence of equipment instability, become significant. With experiments within a narrow range of dilution rates, particularly when looking specifically for values of m, it is more accurate to use a plot of the reciprocal of the ATP yield (g dry wt/mol ATP) vs the reciprocal of the dilution rate which is a straight line intersecting the y axis at $1/Y_{ATP,max}$ (the reciprocal of the ATP yield when $m = 0$). The slope of this line equals m.

Fig. 4 shows a double reciprocal plot for continuous cultures of mouse LS cells. The lower line represents the normal medium, while for the experiments along the upper line, the Na^+ concentration of the medium was increased by 50%. The $1/Y_{ATP,max}$ values at the intercept on the y axis correspond to a $Y_{ATP,max}$ of 40 g dry wt/mol ATP, almost identical to the

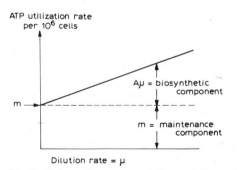

Fig. 3. Schematic representation of the rate of energy utilization for different dilution rates in continuous culture; m = specific maintenance energy coefficient.

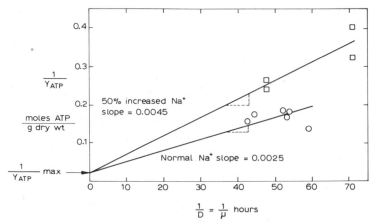

Fig. 4. Plot of the reciprocal of the ATP yield against the reciprocal of the dilution rate for continuous cultures of LS cells grown at normal and 50% increased Na^+ concentration. The slope of these lines = m, the specific maintenance coefficient based here on the dry weight of cells produced. The intercept on the y axis is the reciprocal of the maximum Y_{ATP} when m = 0, and in this case = 40 g dry wt of cells per mol of ATP.

value of 42 determined from batch culture data. The slope of the lower line gives m = 0.0025 mol ATP/g per h or $1.5 \cdot 10^{-11}$ mol ATP/cell per day. This compares well with the value of $1.7 \cdot 10^{-11}$ mol ATP/cell per day found from batch cultures indicating that whatever changes in physiology occur during the course of a batch culture, they have very little effect on the evaluation of m. It can be concluded that our batch procedure for determining m and its variation with external ion concentration, (outlined in Section III) should be as valid as the much more complex technique using continuous culture.

V. AN ESTIMATE OF Na^+ PUMPING COSTS FROM CONTINUOUS CULTURE DATA

From Fig. 4, the change in maintenance energy for a 50% increase in external Na^+ concentration was 0.002 mol ATP/g dry wt per h which is equivalent to $0.7 \cdot 10^{-16}$ mol ATP/cell per s. From data presented by Lamb and MacKinnon [6], it can be estimated that the normal Na^+ flux is $0.4 \cdot 10^{-16}$ mol/s per cell for LS cells. This estimate assumes that the ratio of the Na^+ flux to K^+ flux is identical for L and LS cells (the LS cell is a variant of the L cell, in that it grows spontaneously in suspension) and that the effective area of the LS cell is 2 times its apparent area. If the Na^+ flux is a linear function of the external Na^+ concentration, and if the entire increase in maintenance energy is attributable to Na^+ pumping, this energy should account for 50% of the normal flux = $0.2 \cdot 10^{-16}$ mol/s per cell. On this basis the number of moles of Na^+ expelled per mol of ATP hydrolyzed = 0.29; significantly less than the generally accepted values of 1—3 mol Na^+/mol ATP. However, it should be remembered that the above estimate was made by assuming that the entire increase in m fuels the Na^+ pump. It is not known what other energy demands might be imposed upon a cell by such a drastic increase in its external Na^+ concentration. Based on the value of 0.29 mol Na^+ transported/mol ATP utilized, the Na^+ pump consumes $1.4 \cdot 10^{-16}$ mol ATP/cell per s which corresponds to 82% of the maintenance energy, or 50% of the overall energy available to a cell growing at $\mu = 0.6$ days^{-1}. This estimate seems reasonable, compared to the accepted values of about 20% for giant axons and other cells with surface to volume ratios much less than LS cells, and clearly does not support the proponents of the "Calorific Catastrophe" [7] who fear that there is not enough energy available for any active transport.

REFERENCES

1. Pirt, S.J. (1965) The maintenance energy of bacteria in growing cultures. Proc. Roy Soc. London Ser. B. 163, 224—231.
2. Stouthamer, A.H. and Bettenhausen, C. (1973) Utilization of energy for growth and maintenance in continuous and batch cultures of microorganisms. Biochim. Biophys. Acta 301, 53—70.
3. Kilburn, D.G., Lilly, M.D. and Webb, F.C. (1969) The energetics of mammalian cell growth. J. Cell Sci. 4, 645—654.
4. Paul, J. (1965) In Cells and tissues in culture (Willmer, E.N., ed.) Vol. I, pp. 239—276, Academic Press, New York.
5. Herbert, D., Elsworth, R. and Telling, R.C. (1956) The continuous culture of bacteria; a theoretical and experimental study. J. Gen. Microbiol. 14, 601—622.
6. Lamb, J.F. and MacKinnon, M.G.A. (1971) Effect of ouabain and metabolic inhibitors on the Na^+ and K^+ movements and nucleotide contents of L cells. J. Physiol. London 213, 665—682.
7. Minkoff, L. and Damadian, R. (1973) Caloric catastrophe. Biophys. J. 13, 167—178.

Comparative Physiology — Functional Aspects of Structural Materials
Eds L. Bolis, H.P. Maddrell and K. Schmidt-Nielsen
© North-Holland Publishing Company — 1975 — Amsterdam

Energy utilization for active transport by the squid giant axon

P.C. CALDWELL

Department of Zoology, University of Bristol, Bristol, (U.K.)

GENERAL CONSIDERATIONS

The squid giant axon is one of the few cells for which it is beginning to be possible to draw detailed conclusions about the pattern of energy use. This is particularly so for the active transport processes. The rate of energy use of an axon can be calculated from the rate at which ATP and the phosphagen, arginine phosphate, are used up if the axon is poisoned with cyanide. Values between 3.8 and 6.2 nM of ATP per gram of axoplasm per s can be calculated from the data available [1,2]. These values are obtained from measurements of the ATP and arginine phosphate in axoplasm extracted from the axon at the end of the exposure to cyanide. They represent therefore the rate of ATP utilization by the axon itself and do not include contributions from other cells such as Schwann cells. This rate of ATP utilization is equivalent to an energy consumption of 222 600—363 300 nW/g if the free energy made available by ATP splitting in the axoplasm is 58 590 J/g molecule (equivalent to the 14 000 cal/mol calculated by Caldwell and Schirmer [3]).

This represents only a small part of the total energy consumption of the animal even though two 800 μm hindmost stellar giant axons of 10 cm or more in length contain more than 0.1 g of axoplasm and represent, for single nerve cells, a significant proportion of the total weight of the animal which is about 550 g for a specimen of *Loligo forbesi* with a mantle about 30 cm long. The rates of energy consumption of the fin and mantle muscles, which constitute about 56% of the total weight, are probably very much higher although no values are available. Values in the region of 100—1000 nM of ATP per g per contraction can, however, be calculated from data for muscles from other species (see for example [4,5]).

One of the main energy consuming processes taking place in the squid giant axon is the coupled transport of Na$^+$ and K$^+$ which is needed to maintain the concentration gradients of these ions necessary for the conduction of action potentials. According to the data of Hodgkin and Keynes [6] and of Caldwell and Keynes [7] the active ouabain sensitive efflux is in the region of 40 pM/cm^2 per s. It seems that most of this active efflux is required to counterbalance the largely passive sodium influx which is in the region of 32—49 pM/cm^2 per s [8]. The Na$^+$ entry per impulse is only about 3.5 pM/cm^2 [9] and it seems likely that the giant axons only fire when the animal occasionally contracts the whole mantle in a sudden backwards escape reaction.

The contribution of membrane processes to the total ATP utilisation has been studied by Baker and Shaw [2]. They measured the rate at which ATP and arginine phosphate were used (1) in intact cyanide-treated axons, (2) in membrane-free extruded axoplasm treated with cyanide and (3) in axons treated with ouabain and cyanide. They assumed that the difference between

TABLE I

ENERGY UTILISATION BY THE SQUID GIANT AXON

Data in (A) obtained from Caldwell (1) and Baker and Shaw (2). Na$^+$ efflux from data in Hodgkin and Keynes (6) and Caldwell and Keynes (7). K$^+$ influx from Caldwell et al. (8) Ca^{2+} efflux from Blaustein and Hodgkin (10). Mg^{2+} efflux from Baker and Crawford (12) Glutamate and Glycine influxes from Caldwell, P.C. and Lea, T.J., unpublished. Chloride influx from Keynes [13]. Orthophosphate influx from Caldwell and Lowe (19). Free energy rates calculated from data in Caldwell and Schirmer (3), Caldwell (20) and Deffner (18). Average axon diameter taken as 800 μm.

(A) ATP UTILISATION IN CYANIDE TREATED AXONS

	Rate of ATP utilisation (nM/s per g)	Rate at which free energy is made available (nW/g)
(1) Whole axon	3.8 — 6.2	222 600 — 363 300
(2) Isolated axoplasm	4.3	251 900
Membrane (1—2)	1.9	111 300
Ouabain sensitive component of membrane utilisation	0.9	52 700

(B) ENERGY REQUIREMENTS FOR MEMBRANE TRANSPORT PROCESSES

	Flux (pM/cm^2 per s)	Minimum rate of free energy utilisation (nW/g)
Na$^+$ efflux (ouabain sensitive)	40	24 500
K$^+$ influx	19	~0
Ca^{2+} efflux (Na$^+$ dependent)	0.2	310
Mg^{2+} efflux (Na$^+$ dependent)	0.6	570
Glutamate influx (Na$^+$ dependent)	0.02	15
Chloride influx (active)	13	1900
Orthophosphate influx	0.02	~25
Glycine influx (total)	0.5	
Glycine influx (active)	~0.25	~85

(1) and (2) would represent the rate of ATP utilization by membrane processes. The rate of ouabain sensitive ATP utilization by the membrane was obtained by subtraction of (3) from (1). The results are summarised in Table IA.

It is not possible at present to apportion the various uses to which the energy made available by ATP splitting in the axoplasm is put. Some of it is probably used to transport calcium into organelles such as mitochondria since metabolic inhibitors cause a release of Ca^{2+} into the axoplasm [10]. On the other hand it is possible to draw up a more detailed apportionment of the use of the energy released by ATP splitting by the membrane. This has been done in Table IB. The main identifiable component is the energy requirement for tthe Na^+/K^+ pump. Since this system is ouabain sensitive it must use a major part of the ouabain-sensitive component of the membrane ATP utilization. It uses at least 24 500 nW/g of the 52 700 nW/g available and much of the 28 200 nW/g difference must represent free energy dissipated in driving the pump mechanism. If 3 Na^+ are transported per ATP split this rate of free energy dissipation is likely to be about 14 600 nW/g leaving about 13 600 nW/g for other ouabain-sensitive processes.

A number of transport processes have been established as being linked to Na^+ transport by mechanisms in which Na^+ moving down its electrochemical gradient provides energy for the other transport process. The primary energy source for such processes is the ATP splitting associated with Na^+/K^+ transport. Evidence has been obtained that Ca^{2+} [10,11] and Mg^{2+} [12] fall in this category and the estimated minimum amounts of energy use involved are given in Table IB.

Table IB also shows that certain other transport processes are using small amounts of energy. Active Cl^- transport is insensitive to ouabain and external Na^+ [13] and could contribute to the ouabain-insensitive ATP splitting by the membrane. Active orthophosphate transport is on the other hand ouabain sensitive and presumably contributes to the ouabain-sensitive ATP utilisation by the membrane. The energy requirements of both these processes are, however, small compared with the requirement for Na^+/K^+ transport. Orthophosphate uptake by squid axons is Na^+ dependent (Caldwell, unpublished) and Na^+ dependency has been demonstrated for orthophosphate uptake by rabbit vagus nerve [4].

UPTAKE OF AMINO ACIDS

Recently a start has been made in the study of the uptake of ^{14}C-labelled amino acids into squid giant axons [15—17]. The axon is immersed in a medium containing the labelled amino acid. The uptake is measured either by extrusion and counting of the axoplasm after a period of immersion or by continuous counting in the intact axon with an intracellular glass scintilla-

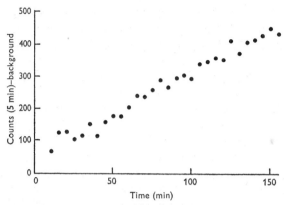

Fig. 1. The uptake of [^{14}C]glycine into a squid axon as measured by an internal glass scintillator. From Caldwell and Lea [17].

tor fibre [17]. In the latter technique a 0.1-mm fibre of a scintillator glass is inserted longitudinally down the axon as centrally as possible. Since the maximum range of the β-particles from ^{14}C is only about 0.3 mm, a glass scintillator fibre in the centre of a 0.8-mm giant axon should only respond to ^{14}C disintegrations occurring intracellularly. The resulting scintillations can be detected with a conventional scintillation counter.

Fig. 1 shows the results of a typical experiment in which the uptake of [^{14}C]glycine was followed continuously. The glycine influx is found to average about 0.5 pM/cm^2 per s. It is not sensitive to the removal of external Na$^+$ and presumably does not derive energy from the Na$^+$ gradient. It appears to consist of two components of roughly equal size. The first is reduced by cyanide and ouabain and is probably an active influx. The energy use, calculated on the glycine gradient for *Loligo* of Deffner [18] is given in Table IB. This aspect of glycine transport could be making use of part of the ouabain-sensitive ATP splitting by the membrane which is not required for Na$^+$ transport, although the actual energy requirement is small. It is not yet possible to assess the overall energy requirement for the amino acids since enough data are not available.

Evidence has been obtained (Caldwell, P.C. and Lea, T.J., unpublished) that the second component of the glycine influx is part of a system involved in the transport of a group of amino acids with similar energy requirements for entry. The most prominent amino acids in the group are glycine, alanine, serine, valine and threonine. Cysteine also probably belongs to this group but no information about the energy requirement is available. The glycine efflux has been studied (Caldwell, P.C. and Lea, T.J., unpublished) and has been found to be substantially increased by the external application of the amino acids just mentioned. This increase in the glycine efflux, like the second component of the influx, is insensitive to cyanide and ouabain. This suggests very strongly that glycine and these other amino acids can enter the axon by an exchange mechanism which derives its energy to a significant extent from the glycine gradient. Glycine would enter in exchange for itself while the other amino acids would enter in exchange for glycine. Information is not yet available to show whether glycine can enter in exchange for the other amino acids or whether they can enter in exchange for themselves.

The glycine efflux is not increased significantly by extracellular glutamate. Glutamate appears to enter the axon mainly by an ATP-dependent process which is reduced by cyanide and the removal of extracellular Na^+ [15,16]. The sensitivity to extracellular Na^+ suggests that glutamate uptake may make use of the Na^+ gradient in contrast to the Na^+-insensitive glycine uptake. Glutamate uptake requires about 15 000 J/mol. If two Na^+ entered with one glutamate enough energy would be made available to transport the glutamate without the intervention of ATP. If on the other hand one Na^+ entered with one glutamate then the energy required from ATP splitting (3000 J/mol) would be about half that required for glycine uptake (6700 J/mol). Such an arrangement might enable glutamate transport to be geared to ATP splitting by the same mechanism as glycine and the requirements for both ATP and external Na^+ could be accounted for. However, this and many other aspects of amino acid and electrolyte transport in the squid giant axon clearly require further investigation before their contribution to the total energy expenditure of the axon can be fully assessed.

REFERENCES

1. Caldwell, P.C. (1960) The phosphorus metabolism of squid axons and its relationship to the active transport of sodium. J. Physiol. London 152, 545—560.
2. Baker, P.F. and Shaw, T.I. (1965) A comparison of the phosphorus metabolism of intact squid nerve with that of the isolated axoplasm and sheath. J. Physiol. London 180, 424—438.
3. Caldwell, P.C. and Schirmer, H. (1965) The free energy available to the sodium pump of squid giant axons and changes in the sodium efflux on removal of the extracellular potassium. J. Physiol. London 181, 25P—26P.
4. Caldwell, P.C. (1953) The separation of the phosphate esters of muscle by paper chromatography. Biochem. J. 55, 458—467.
5. Infante, A.A., Klaupiks, D. and Davies, R.E. (1964) Relation between length of muscle and breakdown of phosphorylcreatine in isometric tetanic contractions. Nature 201, 620.
6. Hodgkin, A.L. and Keynes, R.D. (1955) Active transport of cations in giant axons from Sepia and Loligo. J. Physiol. London 128, 28—60.
7. Caldwell, P.C. and Keynes, R.D. (1959) The effect of ouabain on the efflux of sodium from a squid giant axon. J. Physiol. London 148, 8P—9P.
8. Caldwell, P.C., Hodgkin, A.L., Keynes, R.D. and Shaw, T.I. (1960) The effects of injecting energy-rich phosphate compounds on the active transport of ions in the giant axons of Loligo. J. Physiol. London 152, 561—590.
9. Keynes, R.D. and Lewis, P.R. (1951) The sodium and potassium content of cephalopod nerve fibres. J. Physiol. London 114, 151—182.
10. Blaustein, M.P. and Hodgkin, A.L. (1969) The effect of cyanide on the efflux of calcium from squid axons. J. Physiol. London 200, 497—527.

248

11. Baker, P.F., Blaustein, M.P., Hodgkin, A.L. and Steinhardt, R.A. (1969) The influence of calcium on sodium efflux in squid axons. J. Physiol. London 200, 431—458.
12. Baker, P.F. and Crawford, A.C. (1971) Sodium-dependent transport of magnesium ions in giant axons of *Loligo forbesi*. J. Physiol. London 216, 38P—40P.
13. Keynes, R.D. (1963) Chloride in the squid giant axon. J. Physiol. London 169, 690—705.
14. Anner, B., Ferrero, J., Jirounek, P. and Straub, R.W. (1973) Inhibition of intracellular orthophosphate uptake in rabbit vagus nerve by Na withdrawal and low temperature. J. Physiol. London 232, 47P—48P.
15. Baker, P.F. and Potashner, S.J. (1973) Glutamate transport in invertebrate nerve: the relative importance of ions and metabolic energy. J. Physiol. London 232, 26P—27P.
16. Baker, P.F. and Potashner, S.J. (1973) The role of metabolic energy in the transport of glutamate by invertebrate nerve. Biochim. Biophys. Acta 318, 123—139.
17. Caldwell, P.C. and Lea, T.J. (1973) Use of an intracellular glass scintillator for the continuous measurement of the uptake of ^{14}C-labelled glycine into squid giat axons. J. Physiol. London 232, 4P—5P.
18. Deffner, G.G.J. (1961) The dialyzable free organic constituents of squid blood; a comparison with nerve axoplasm. Biochim. Biophys. Acta 47, 378—388.
19. Caldwell, P.C. and Lowe, A.E. (1970) The influx of orthophosphate into squid giant axons. J. Physiol. London 207, 271—280.
20. Caldwell, P.C. (1969) Energy relationships and the active transport of ions. Current topics in Bioenergetics 3, 251—278.

Comparative Physiology — Functional Aspects of Structural Materials
Eds L. Bolis, H.P. Maddrell and K. Schmidt-Nielsen
© North-Holland Publishing Company — 1975 — Amsterdam

Transport of inorganic phosphates across nerve membranes

R.W. STRAUB, BEATRICE ANNER, J. FERRERO and P. JIROUNEK

Département de Pharmacologie, Ecole de Médecine, CH-1211 Geneve 4 (Switzerland)

INTRODUCTION

Phosphates are known to be important components of biologocal membranes and structural elements of the animal body and to be involved in many biochemical reactions, particularly those connected with metabolism and the transfer of energy. Despite this, relatively little is known on the way by which inorganic phosphates are taken up by animal cells. The present paper reports results of our studies in nerve fibres, as well as observations by other authors in various tissues. An attempt will be made also to estimate the amount of energy needed for cells to keep their internal physiological inorganic phosphate level.

In the body, the major part of inorganic phosphate is found in the form of orthophosphate that, at physiological pH, is present in two ionic forms, about 20% as the mono-valent, the rest as the divalent ion. Both forms are often referred to as inorganic phosphate or P_i.

We have studied the uptake and release of inorganic phosphate in rabbit vagus nerve, a preparation that contains a large number of small non-myelinated nerve fibres [1]. These fibres, often called C-fibres, are known for their particularly high surface to volume ratio [2], the mean diameter of the axons being around 0.7 μm [3]. Transmembranal phosphate fluxes are therefore relatively pronounced in this preparation and can easily be measured with ^{32}P-labelled phosphate.

MATERIALS AND METHODS

The influx measurements were based on two different methods: a method by which the total amount of radioactive phosphate in the preparation was continuously measured, and a method for studying the localization of the labelled phosphate. For the phosphate uptake studies, a desheathed cervical vagus was mounted in a polythene tube, where it was perfused at $37°$ C by Locke medium (containing in mM: NaCl 154; KCl 5.6; $CaCl_2$ 0.9; $MgCl_2$ 0.5; glucose 5.5; Tris 10; pH adjusted to 7.40 with HCl and, in addition, a variable amount of a 0.1 M solution with 20% Na_2HPO_4 and 80% NaH_2PO_4), or ^{32}P-labelled Locke medium, and where the total radioactivity of the preparation and the surrounding fluid was counted and registered [7]. In this way, it was possible to follow the immediate effects of changes in the external fluid on the phosphate influx.

The second method consisted in studying the incorporation of the ^{32}P in various compounds. After washing for 5 min in ice-cold label-free Locke medium, the preparation was removed from the apparatus, plunged into boiling triethanolamine buffer, rapidly cooled, homogenized and de-proteinized with chloroform. After centrifugation, the compounds in the supernatant were separated by thin-layer [4] or column chromatography [5], and the ^{32}P in the different fractions was counted. The total amount of inorganic phosphate, ATP, ADP, and CrP could also be measured, when column chromatography was used.

The efflux was measured by washing the preparation with label-free Locke medium and counting the radiophosphate in the effluent of the apparatus.

RESULTS AND DISCUSSION

Fig. 1 shows the uptake recorded during incubation in 2 mM phosphate Locke medium. In order to avoid, as far as possible, net fluxes of phosphate, the desheathed preparation was equilibrated in unlabelled 2 mM phosphate

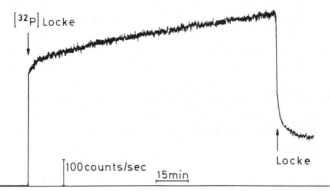

Fig. 1. Uptake of labelled-phosphate by desheated vagus nerve. Record of radioactivity of preparation and surrounding medium shows filling of tube and extracellular space after application of labelled Locke, followed by slow uptake into nerve fibres. Return to tracer-free Locke is followed by loss of radioactivity from tube and extracellular space, and slower efflux from preparation. Locke contained 2 mM phosphate, 100 cps correspond to uptake of 400 μmol phosphate per kg wet wt, temperature $37°$ C.

Locke medium for 1 h before the labelled Locke medium was applied. The record shows that after the rapid filling of the tube and the extracellular space, which in the conditions of this experiment was complete within less than 2 min (see [7]), there was a slow uptake of labelled phosphate by the preparation. Unpublished experiments show that this uptake can be divided into two exponentials, one with a time constant of 22 min, another with a time constant of 11 h at 2 mM extracellular phosphate. The total amount of phosphate in the first fraction is then about 1/10 of that in the second one. The two fractions can be separated also by a differential effect of cations on the labelling of ATP. The experiments described below deal mainly with the second fraction that almost certainly corresponds to intracellular uptake of phosphate.

INFLUX AND EFFLUX OF PHOSPHATE

From records like that shown in Fig. 1 and corresponding efflux measurements, the phosphate influx can be calculated, using conventional methods for analyzing the results (e.g. [6]). Such experiments show that the phosphate influx from Locke medium contains a large saturable component: for extracellular phosphate concentrations between 0.2 and 5 mM, the apparent K_m of this component is around 1 mM [7] and the maximal influx 0.015 mM per kg wet wt/min. Recent experiments showed that the K_m is considerably lower for concentrations between 0.02 and 0.2 mM and may then be as low as 0.02 mM.

Extraction after 40-min incubation revealed the presence of labelled orthophosphate, ATP, ADP, GTP, [7] CrP (unpublished results). The specific activities of the compounds were remarkably low: during incubation in 0.2 mM phosphate, the specific activities of intracellular orthophosphate and ADP increased at the same rate reaching about 10% after 120 min incubation; ATP was then labelled at 20% (counting 1 label per molecule). These observations suggest that the influx of phosphate is relatively slow compared to its metabolic turnover. The latter can be estimated also from the O_2 consumption [17] and the total amount of ATP in this preparation [8,9]. It is then evident that it takes about 5 min for the total amount of ATP to be used and re-synthesized. This would correspond to a metabolic turnover of 0.36 mmol P/kg wet wt per min; on the other hand, the maximal saturable influx is, as mentioned, around 0.015 mmol/kg wet wt per min, and at the extracellular phosphate concentration found in vivo, the total influx is about 0.005 mmol/kg per/min. Studies with labelled orthophosphate injected into squid axons show that similarly the turnover of intracellular phosphate is much faster than the efflux or influx (cf. [10—12]).

The intracellular concentration of orthophosphate measured in the vagus is around 5 mM (see [9]) the in vivo concentration of phosphate outside the

252

axons is not exactly known, but probably not very different from the 0.5 mM found, for instance, in human cerebro-spinal fluid [13], or the 1 mM of the thoracic duct [14]. With a resting potential of -34 mV [3], the distribution of mono- and divalent phosphate is therefore far from its electrochemical equilibrium, and the high intracellular concentration can be maintained only by the working of an uphill transport system that causes an inward flux.

Mullins [15] showed that in frog sciatic nerve the uptake of phosphate is blocked by O_2 lack, low temperature and azide; in squid axon, Caldwell and Lowe [12] found that the influx of inorganic phosphate is greatly decreased by metabolic inhibitors and by ouabain. The observations suggested the presence of an active, ouabain-sensitive transport process. However, in rabbit vagus, Anner [16] observed only a 25% decrease in influx with ouabain and a 50% decrease with metabolic inhibitors after 1 h incubation. The effect of ouabain was maximal with 10^{-4} M, no larger effects were found with higher concentration.

Since many transport processes somehow depend on the presence of extracellular Na^+, it appeared interesting to study the effect of this ion on the phosphate influx. Fig. 2 shows an experiment [7] where the $^{32}P_i$ uptake was recorded: replacement of extracellular Na^+ by choline greatly slowed the influx. The decreased influx was maintained as long as Na^+ was absent. When Na^+ was re-introduced, the influx recovered immediately. Since some

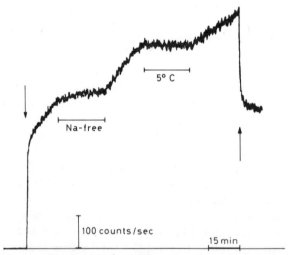

Fig. 2. Effects of Na^+ withdrawal and low temperature on uptake of radiophosphate in desheathed rabbit vagus. Record of phosphate uptake by preparation shows slowing of influx when extracellular Na^+ is replaced by choline and abolition of influx, when temperature is lowered, both effects being reversible; arrows indicate application and withdrawal of labelled solutions. Locke contained 0.2 mM phosphate, 100 cps correspond to uptake of 40 μmol phosphate/kg wet wt, temperature, unless otherwise stated, 37°C.

influx continued in the absence of sodium, the phosphate influx contains a Na^+-sensitive part (about 80—90% of the total influx in 0.3—3 mM extracellular phosphate) and a Na^+-insensitive one. When, in other experiments, the incorporation of ^{32}P into ATP, ADP and GTP was measured, it was found that during 40 min incubation in Na^+-free (choline) Locke medium the incorporation was reduced by about 70% [7]. Thus, the results of three types of experiments, viz. the measurement of influx at different extracellular phosphate concentrations, the partial dependence of influx on extracellular Na^+, and the effect of Na^+ on the incorporation of radioactive phosphate into ATP, are consistent with the idea that the phosphate influx can be divided into a Na^+-dependent, saturable component, and a Na^+-insensitive, non-saturable one.

Both parts appear to be highly influenced by temperature, as illustrated in Fig. 2, where the lowering of the temperature from 37 to 5° C abolished the uptake process. Experiments on the incorporation showed that after 40 min incubation at this temperature, no labelled nucleotides could be detected [7] although metabolism, as judged from the O_2 consumption [17] was not completely inhibited.

At extracellular concentrations of 0.2—1 mM replacement of Na^+ by Tris similarly reduced the phosphate uptake; when K^+ was used, the influx was decreased by 50%. A somewhat similar effect of Na^+ on phosphate influx has been observed by Chambers [18] in fertilized sea urchin eggs and by Harrison and Harrison [19] in rat intestine, and Abood [20] showed that in bullfrog spinal ganglia partial replacement of Na^+ by K^+ decreased the uptake of radiophosphate and its incorporation into ATP and CrP.

Na ions could, by their electrochemical gradient, provide the driving force for the phosphate influx; on the other hand, it is possible that their presence in the extracellular fluid, in some way, is necessary for the maintenance of influx, for instance by neutralizing a negative charge on a phosphate carrier complex or by catalyzing a biochemical reaction.

Several studies were undertaken to analyze the mode of action of extracellular Na^+. First the effect of different extracellular Na^+ concentrations was tested. These experiments showed that there was little decrease in influx until the extracellular Na^+ was lowered below 40 mM. With an intracellular Na^+ of 60 mM and a resting potential of −34 mV this would correspond to a 60% decrease of the electrochemical gradient for Na^+. The second series of experiments consisted in adding KCl (150 mM) to Locke medium, a procedure by which the resting potential is almost completely abolished. In this condition, there was no detectable change in the rate of phosphate influx, for extracellular phosphate concentrations from 0.2 to 0.6 mM. If the Na^+ sensitive influx is driven by the electrochemical potential for Na^+, the influx through this mechanism should have strongly decreased, since the gradient was lowered by more than 40%. On the other hand, the phosphate influx that is insensitive to Na^+, and that is probably dependent on the electro-

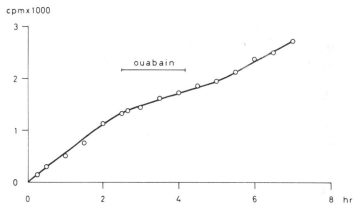

Fig. 3. Effect of ouabain on uptake of radiophosphate by desheathed rabbit vagus. Plot of radioactivity of preparation from experiment like that shown in Fig. 1 shows slowing of influx when ouabain (0.2 mM) is applied, with partial recovery after removal of drug. Locke contained 0.08 mM phosphate, 1000 cpm correspond to uptake of 300 μmol/kg wet wt, temp. 37°C.

chemical potential for phosphate (it is linearly related to the extracellular phosphate concentration) would greatly increase so that the sum of both effects might well result in little change in phosphate uptake. As mentioned, replacement of Na^+ by K^+ results in a decrease in influx to about 50% of that in Locke medium, which is in good agreement with this hypothesis. Since the relation between Na^+ sensitive and insensitive influx is known, as well as the extra- and intracellular ion concentrations and the membrane potential, the argument can be pursued quantitatively, assuming that the Na^+-sensitive influx depends on a positively charged complex, the Na^+-insensitive one on the negatively charged phosphate. Results of such calculations agree with the effects observed after adding KCl or replacing Na^+ by K^+.

The third experiment consisted in applying ouabain [21] at such a concentration that the intracellular Na^+ increases almost to the extent of the extracellular one, with a corresponding lowering of K^+ [22], while maintaining a normal ATP content [23]. This procedure also abolished the resting potential, since both the K^+ diffusion potential and the working of the electrogenic Na^+ pump [24] are affected. Fig. 3 shows that ouabain caused only a small decrease in phosphate influx, which again is understandable, if the Na^+-sensitive influx is abolished, while the Na^+-insensitive influx is increased.

The decrease in phosphate influx by metabolic inhibitors [16] is more difficult to interpret, since at the concentrations used, these agents also lower the ATP content, cause an increase in Na^+ and a K^+ loss [23], and also raise the intracellular ortho-phosphate (Schorderet, M., unpublished results).

MEMBRANE FRACTION

As mentioned, the uptake of phosphate by the preparation can be divided into two fractions: a fraction of relatively rapid uptake, and a much slower one. The first fraction shows a time constant of about 20 min, that is independent of extracellular phosphate; the time constant of the second fraction is much longer and varies, because of the saturation kinetics of the Na^+-sensitive part, with extracellular phosphate. Another difference comes from the effect of external cations. While a large proportion of the slow fraction depends on the presence of extracellular Na^+, which for this action cannot be replaced by choline or Tris, the first fraction is also influenced by cations, but in this case the uptake is maintained by Na^+ and choline, and abolished by K^+. It is not clear as to which structural components the two fractions can be assigned to. The possibility that one of the fractions corresponds to the uptake into myelinated fibres of the cervical vagus can, however, be excluded, since both fractions are also found in the thorical vagus, where, in the rabbit, myelinated fibres are almost completely absent [1].

Uptake during the first fraction is not accompanied by incorporation of labelled-phosphate into nucleotides. This fraction therefore appears to be situated in the membrane rather than in the axoplasm or the cytoplasm of the Schwann cells. This part of phosphate uptake shows some similarity with the membrane fraction described in frog striated muscle by Causey and Harris [25].

ENERGY REQUIREMENT

For the estimation of the energy requirement the first, rapid fraction of phosphate uptake can probably be neglected: it is relatively small compared to the second, slow fraction and probably located in the membrane rather than inside the cells. The maintenance of the intracellular fraction, however, requires energy, since it depends on uphill fluxes driven by the electrochemical potential for Na^+; it therefore finally depends on the working of the Na^+ pump.

In conditions ressembling those in vivo, i.e. in a nerve equilibrated with 0.5 mM extracellular phosphate the slow uptake amounts to 0.008 mmol/kg wet wt per min, 80% of which is Na^+-sensitive, so that the true uphill flux is 0.006 mmol/kg wt per min. The Na^+ fluxes in the steady-state have not been measured in this preparation, but almost certainly they are not very different from the K^+ fluxes, which have been studied in detail by Keynes and Ritchie [3], and amount from their Tables 2 and 3, to 1.2 mmol/kg wet wt per min at $37°C$. The maintenance of the phosphate balance therefore requires about 0.5% of the Na^+ flux. The resting O_2 consumption, which gives some measure of the total energy expenditure in this preparation is decreased by 26%

when ouabain is applied [17], indicating that about 1/4 of the total energy is used for the Na^+/K^+ pump. In conclusion then, the maintenance of the phosphate balance would require about 0.125% of the total energy consumed by a resting vagus nerve.

In squid axons the phosphate fluxes are very similar to those in the vagus, if calculated on the basis of $1 \, cm^2$ membrane. A phosphate influx of 21 $fmol/cm^2$ per s has been found by Caldwell and Lowe [12]. If this flux similarly depends on the Na^+ gradient, it would represent 0.12% of the Na^+ influx of 17 $pmol/cm^2$ per s [26,27]. This percentage is close to that found in vagus, suggesting that widely different animals might use about the same fraction of their active Na^+ transport for the phosphate balance.

The percentage of the total energy devoted to phosphate transport might well be more variable, if different cells are compared: because the rabbit vagus nerve fibres are exceedingly thin, the fraction of total energy required for the phosphate balance probably represents an upper limit rather than a mean, which in most cells may even be smaller. This of course reflects the exceedingly low passive phosphate permeability that has been developed by many animal cells, to prevent the escape of an important anion, phosphate.

ACKNOWLEDGEMENT

This study was supported by S.N.S.F. Grants 3.286.69 and 3.0890.73.

REFERENCES

1. Evans, D.H. and Murray, J.G. (1954) Histological and functional studies on the fibre composition of the vagus nerve of the rabbit. J. Anat. Lond. 88, 320—337.
2. Ritchie, J.M. and Straub, R.W. (1957) The hyperpolarization which follows activity in mamalian non-medullated fibres. J. Physiol. London 136, 80—97.
3. Keynes, R.D. and Ritchie, J.M. (1965) The movement of labelled ions in mammalian non-myelinated nerve fibres. J. Physiol. London 179, 333—367.
4. Randerath, E. and Randerath, H. (1964) Resolution of complex nucleotic mixture by two dimensional anion exchange thinlayer chromatography. J. Chromatogr. 16, 126—129.
5. Garrahan, P.J. and Glynn, J.M. (1967) The incorporation of inorganic phosphate into adenosine triphosphate by reversal of the sodium pump. J. Physiol. London 192, 237—256.
6. Hodgkin, A.L. (1951) The ionic basis of electrical activity in nerve and muscle. Biol. Rev. 26, 339—409.
7. Anner, B., Ferrero, J., Jirounek, P. and Straub, R.W. (1973) Inhibition of intracellular orthophosphate uptake in rabbit vagus nerve by Na^+ withdrawal and low temperature. J. Physiol. 232, 47P—48P.
8. Greengard, P. and Straub, R.W. (1959) Effect of frequency of electrical stimulation on the concentration of intermediary metabolites in mammalian non-myelinated nerve fibres. J. Physiol. London 148, 353—361.

9. Chmouliovsky, M., Schorderet, M. and Straub, R.W. (1969) Effect of electrical activity on the concentration of phosphorylated metabolites and inorganic phosphate in mammalian non-myelinated nerve fibres. J. Physiol. London 202, 90P—92P.

10. Tasaki, I., Teorell, T. and Spiropoulos, C.S. (1961) Movement of radioactive tracers across squid axon membrane. Am. J. Physiol. 200, 11—22.

11. Caldwell, P.C., Hodgkin, A.L., Keynes, R.D. and Shaw, T.I. (1964) The rate of formation and turnover of phosphorus compounds in squid giant axons. J. Physiol. London 171, 119—131.

12. Caldwell, P.C. and Lowe, A.G. (1970) The influx of orthophosphate into squid giant axons. J. Physiol. London 206, 271—280.

13. Friedman, A. and Levinson, A. (1955) Cerebrospinal fluid inorganic phosphorus in normal and pathologic conditions. Arch. Neurol. Psychiat. (Chic.) 74, 424—440.

14. Werner, B. (1966) The biochemical composition of the human thoracic duct lymph. Acta Chir. Scand. 132, 63—76.

15. Mullins, L.J. (1954) Phosphate exchange in nerve. J. Cell. Comp. Physiol. 44, 77—86.

16. Anner, B. (1973) Localisation, répartition et transport de l'ion phosphate dans un tissu nerveux. Thèse No. 1609, pp. 1—66, Fac. Sciences, Univ. Genève, Editions Médecine et Hygiène, Genève.

17. Ritchie, J.M. (1967) The oxygen consumption of mammalian non-myelinated nerve fibres at rest and during activity. J. Physiol. London 188, 309—329.

18. Chambers, E.L. (1963) Role of cations in phosphate transport by fertilized sea urchin eggs. Fed. Proc. 22, 331.

19. Harrison, H.E. and Harrison, H.C. (1963) Sodium, potassium and intestinal transport of glucose, L-tyrosine, phosphate and calcium. Am. J. Physiol. 205, 107—111.

20. Abood, L.G. (1968) Interrelationships between phosphates and calcium in bioelectric phenomena. Int. Rev. Neurobiol. 9, 223—261.

21. Anner, B., Ferrero, J., Jirounek, P. and Straub, R.W. (1973) Na^+-dependent phosphate influx into mammalian nerve fibres. Experientia 29, 740.

22. Wespi, H.H., Mevissen, D. and Straub, R.W. (1969) The effect of ouabain and ouabagenin on active transport of sodium and potassium in vagus nerve fibres. Arch. Int. Pharmacodyn. 181, 307—315.

23. Chmouliovsky, M. and Straub, R.W. (1974) Increase in ATP by reversal of the Na^+-K^+-pump in mammalian non-myelinated nerve fibres. Pflüg. Arch. in the press.

24. Rang, H.P. and Ritchie, J.M. (1968) On the electrogenic sodium pump in mammalian non-myelinated nerve fibres and its activation by various external cations. J. Physiol. London 196, 183—221.

25. Causey, G. and Harris, E.J. (1951) The uptake and loss of phosphate by frog muscle. Biochem. J. 49, 176—183.

26. Caldwell, P.C., Hodgkin, A.L., Keynes, R.D. and Shaw, T.I. (1960) The effects of injecting energy-rich phosphate compounds on the active transport of ions in the giant axons of Loligo. J. Physiol. London 152, 561—590.

27. Hurlbut, W.P. (1970) Ion movements in nerve. In membranes and ion transport (Bittar, E.E. ed.), pp. 95—143, Wiley-Interscience, London, New York, Sidney, Toronto.

Comparative Physiology — Functional Aspects of Structural Materials
Eds L. Bolis, H.P. Maddrell and K. Schmidt-Nielsen
© North-Holland Publishing Company — 1975 — Amsterdam

Energetics of sugar transport

WALTER WILBRANDT

Department of Pharmacology, University of Bern, Bern (Switzerland)

SUMMARY

The energetics of active sugar transport are characterized by a multitude of different coupling mechanisms. In most interpretations the mechanism involves a carrier system linked energetically either to chemical reactions or to flow of matter. Reaction coupling may occur by reversible changes in carrier-substrate affinity, due to phosphorylation or oxydation of the carrier by reactions with metabolites. Flow coupling rests on common elements (e.g. common carrier) for two transport systems. It includes the important group of Na^+-dependent transport systems.

An entirely different mechanism not involving mobile carries at all appears to be constituted by the phosphotransferase systems in which transport and enzymatic phosphorylation appear to be so closely related that they may be considered to be identical. The transport in this case is interpreted in terms of group translocation, one form of Mitchell's general principle of chemiosmotic coupling.

INTRODUCTION

Sugar is transported across biological membranes by a variety of different mechanisms. They can, with a certain degree of arbitrariness, grouped as follows.

(1) Carrier mediated transport (facilitated transport)

(2) Active transport (against gradients of chemical potential) deriving energy from metabolism.

(3) Active transport, deriving energy from the flow of other matter: flow coupled transport.

(4) Active transport, using the energy of one single enzymatic reaction (chemiosmotic coupling of Mitchell).

The seemingly most natural mechanism, namely simple diffusion, is not listed in this categorisation. Indeed it appears to be quite rare. Living cells seem, in general, to be impermeable to sugar (disregarding carrier mediation), a fact which struck early workers in the field of permeability and induced a number of speculations. The largest group is the first. It comprises muscle cells, heart muscle cells, brain cells, liver cells, red blood cells, tumor cells, in fact the majority of cell types. In contrast to amino acids active uptake is an exception rather than the rule. It is limited to epithelial cells in kidney and intestine in mammals and to bacterial cells.

CARRIER MEDIATION

Although the first group does not require an energy supply, a short discussion of the mechanism appears useful. Carrier mediated transport systems have mainly been studied in erythrocytes [1]. Only the red cells of man and ape have this system with a considerable capacity. Some studies have been performed in rabbit cells in which, however, the mediated transport is slow.

Various observations have led to the concept that in these cells the membrane as such is impermeable for sugar, that some sugars, however, are enabled to pass the membrane by reaction with a membrane component, the carrier, to form a carrier-substrate complex capable of membrane passage. On the other side ("trans side") of the membrane the free substrate is released to the internal medium. Under certain simplifying conditions the kinetics of the transport can be described by relatively simple equations [2]. Within limits they resemble those of enzymatic reactions, particularly in the dominating role of two parameters: V, the maximum rate, and K_m, the Michaelis constant (dissociation constant of the transport complex).

Interesting consequences of carrier mediation emerge under the condition of high carrier saturation on both sides of the membrane. They include biphasic dependence of transport rate on substrate concentration and on affinity (decreasing rate with increasing concentration or affinity in the range of high saturation) and particularly the interdependence of simultaneous movements of two substrates across the system described as counter transport and competitive acceleration.

The system of carrier mediation described has no energy input and is not capable of doing work, i.e. transporting substrate against gradients. It leads to equilibration of the concentrations on the two sides of the membrane and then the net movement ceases. Yet, many active transport systems operating uphill are currently interpreted in terms of a carrier mechanism. Also, in many of them kinetic characteristics of carrier mediation have been reported. The question, thus, arises, how carrier mediated transport can be enabled to operate uphill, in other words, what is necessary to transform an equilibrating carrier system into a carrier pump.

The general answer is that energy yielding processes must in a suitable manner be coupled to the carrier system [3]. If the energy yielding process is a metabolic reaction the coupling can occur by chemical reaction between metabolites and components of the carrier system. If the energy comes from simultaneous flow of matter the system responsible for this flow must have common elements with the carrier system of the actively transported substrate.

An example of reaction coupling is the coupling to metabolic reactions as visualized by Rosenberg [4] 10 years ago. His CZ-system is a carrier system in which the carrier is able to assume either one of two interconvertible forms, C and Z, with high and low affinity to the transport substrate, respectively. A pump action ensues if, due to reactions between carrier and metabolites, the concentration ratio Z/C is unequal on the two sides of the membrane. Then, if the concentration of C (high affinity form) is large on the external and low at the internal surface of the membrane, much transport substrate is taken up by the carrier from the external medium, in spite of low substrate concentration and little from the interior, in spite of high substrate concentration. Thus accumulation emerges.

The kinetics have been worked out by Rosenberg who showed that the rate equation can be written in a form similar to that of the equilibrating (not pumping) carrier system except that K_m is replaced by terms different on the two sides of the membrane. They involve, at least on one side of the membrane, concentrations and parameters of the carrier-metabolite reaction.

Cases of flow coupling include various forms of counter transport and co-transport, particularly, according to current interpretation, a number of Na^+-dependent systems. The necessary condition for flow coupling is a common element in the two transport systems involved, mostly the carrier.

REACTION COUPLED SYSTEMS

The most important observations on active sugar transport across cell membranes linked to metabolism, concern bacterial cells. The discovery of bacterial permease systems some 30 years ago, summarized brillantly by Cohen and Monod [5] has opened a wide field of fascinating studies. They involve a considerable number of different cell types as well as of substrates. Substrates are mainly sugars, amino acids and other anions. The uptake is against gradients leading to accumulation ratios up to 100—1000 and depending on the metabolism of the cell as indicated by blocking action of uncoupling agents like dinitrophenol, azide and others or, in other cases, by inhibitors of the electron transfer chain like cyanide, sulfhydryl reactors and others. A most important tool for the study of these systems is provided by their genetic control.

A number of systems have been recognized, differing in the type of energetic coupling.

The Galactoside Permease System

The system allowing uptake of lactose and other galactosides in anaerobic organisms including *E. coli* has been analyzed extensively [6]. While the natural substrate lactose is metabolised inside the cell rapidly, beginning with hydrolysis by galactosidase, such that intracellular accumulation is not conspicuous, a number of non-metabolizable derivatives, the thiogalactosides, have been successfully used to analyse the mechanism of accumulation. A case of competitive acceleration has been described by Kepes: uptake of thiogalactoside is inhibited competitively by higher concentrations of thiomethylgalactoside while the lowest concentration of this competitor has an accelerative action. This is in excellent accordance with kinetical prediction for carrier mediation. Also counter transport has been demonstrated repeatedly. This was the main reason to abandon the original concept that only uptake uses a transport system while exit occurs by diffusion.

With respect to energy coupling findings of Winkler and Wilson [7] were of importance. Energy coupling can be eliminated by inhibitors or by lack of metabolic substrate. Furthermore, mutants were found lacking the coupling mechanism and, therefore, being unable of substrate accumulation. These cells were shown not to have lost the ability of taking up sugar entirely, but only the capacity to accumulate it against gradients. The transport system, then, has properties of equilibrating carrier mediated transport exhibiting counter transport in a conspicuous way. The transport constants K_t inward and outward become equal but not by a decrease of the affinity (rise of K_t) for entrance but on the contrary by an increase of affinity (decrease of K_t) for outward transport. Thus, apparently the coupling to metabolism, contrary to possible expectation, acts by inhibiting outflow rather than by activating inflow.

The mode of action of this affinity change has not been analysed in the case of the galactoside system. The inhibitory effect of uncouplers might be taken as indication for phosphorylation. At any rate, however, this mechanism cannot be considered to be general as the observations of Kaback [8] on the effect of D-lactate dehydrogenase show.

The D-Lactate Dehydrogenase System

Kaback [8] discovered that a considerable number of bacterial transport systems appear to derive energy from the electron transfer rather than from high energy phosphate. He found that in *Escherichia coli* membrane vesicles the transport of 15 amino acids and 6 sugars is dramatically stimulated by the addition of D-lactate and inhibited by cyanide and other inhibitors of the electron transfer chain, not, however, by dinitrophenol and other uncouplers. He visualizes in these cases the carriers as intermediates of electron transfer and the change in affinity as brought about by oxidation and reduction, the oxidized form having high affinity, the reduced form low affinity.

Flow Dependent Systems

In a number of cases of active sugar transport it appears that the necessary energy is not derived directly from metabolism but rather from the simultaneous movement of other transport substrates across the same system. To this group belong in the first place the now numberous cases of Na^+ dependence. They are interpreted in terms of carriers with at least two separate binding sites for transport substrates and for Na^+. The movement of Na^+ can induce or modify the movement of transport substrate either by co-transport or by counter-transport.

The first pertinent observation was made by Riklis and Quastel [9] showing that the intestinal absorption of sugar depends on the presence of Na^+. This observation was confirmed and extended in a number of laboratories. Na^+ dependence was shown for intestinal absorption of sugars and amino acids as well as for their uptake into tubular cells in kidney. A number of other substrates was shown to exhibit Na^+ dependence in a similar way.

The interpretation in terms of ternary transport complexes involving carrier, substrate and Na^+, was first given by Crane [10] and worked out kinetically in great detail by Schultz and Curran [11]. The most frequent case is an effect of Na^+ on the affinity rather than on the mobility of sugar transport. A velocity type, however, was also observed in one case: in the sugar absorption from rabbit intestine. Here Na^+ affects V and not K_m.

In accordance with kinetical prediction the effect was shown to be mutual, i.e. not only does Na^+ accelerate the transport of sugar but also vice versa Na^+ transport is enhanced by sugar.

The Phosphotransferase System

An interesting group of observations has recently been made indicating a close relationship between phosphorylation of a number of sugars and their transport across the bacterial cell membrane. The observations differ from those discussed so far in different respects: there is no indication for the involvement of a mobile carrier and the transmembrane transport is closely related to intramembranal enzymatic reactions leading to phosphorylation of the sugar during the passage across the membrane.

The first observations were of a purely biochemical nature. A bacterial phosphotransferase system was found in *E. coli*, *S. typhimurium*, *B. subtilis* and *S. aureus* [12]. The system transfers phosphate from phospho-enol pyruvate to sugars in a chain of consecutive reactions. In a first step a low molecular weight protein (HPr) is phosphorylated with the help of a first enzyme I. In a second step then HPr, by means of a second enzyme II, phosphorylates a number of sugars. The first two proteins involved, enzyme I and HPr, are water soluble and their action is unspecific with respect to sugars. Enzyme II (which can consist of two components) appears to be membrane bound and to have specificity for the sugars involved. The overall effect, thus, is the transfer of phosphate from phosphoenol pyruvate to sugars and requires the presence of at least three proteins, enzymes I, HPr and enzyme II. Lack of enzyme I or HPr abolishes the phosphorylation which can readily be reactivated by the addition of the lacking protein.

Later it was found that not only the phosphorylation of sugars depended on the proteins described but also their uptake. Furthermore, the sugars appeared in the cell not in the free form but as phosphates. Thus, the phosphorylation appeared to be closely related to membrane transfer.

The question arose whether phosphorylation takes place before or immediately after passage of the diffusion barrier in the membrane. Both possibilities could be ruled out in a convincing way. Sugar 6-phosphates are poor penetrants, disproving phosphorylation previous to membrane passage. Elegant experiments with double radioisotope labelling techniques provided convincing evidence that phosphorylation after passing the membrane can also not underly the appearance of sugar phosphate in the cells. Thus, the conclusion appeared inevitable that phosphorylation does not occur prior to or subsequent to membrane passage but concommittantly with it.

The mechanism, therefore, was interpreted in terms of group translocation, a concept introduced by Mitchell [13] about 15 years ago. Mitchell vizualizes a membrane bound enzyme with sided accessibility to the participants of the enzymatic reaction. In one of his schemes the enzymatic reaction consists in the transfer of the group G from D to A. The basic assumption is the intramembraneous existance of an enzyme catalyzing this transfer. The active site is accessible to DG only from one side of the membrane and to A and AG only from the other side. Mitchell visualizes the molecular topography such that membrane passage of G cannot occur except across the active center of the enzyme. In this case then the energy for active transport would be the free energy of the enzymatic reaction. Reaction and transport could be described in the same terms. This would certainly constitute a most simple possible way of energetic coupling. Mitchell's views have been most stimulating and discussion provoking in fields not belonging to the scope of this review. Particularly, the application of his principle of chemiosmotic coupling to the problem of oxidative phosphorylation has led to very lively discussion.

THE ENERGETIC COST OF SUGAR TRANSPORT FOR MAMMALIAN ORGANISMS

It was pointed out that membrane passage of sugars in most cells and tissues occurs by carrier mediation without energy coupling. The transport of sugar across cell membranes by such systems requires no energy. Thus, in contradistinction to amino acids, entry into most cells requiring sugars creates no energy needs.

The only important exception are the transporting epithelia in intestine and kidney. In these cells sugar transport is active, operating against gradients of chemical potential. However, according to current ideas, the systems involved are prevailingly systems flow coupled to Na^+ movements along Na^+ gradients. The energy stores of these gradients are maintained by the Na^+-K^+ pumps. The term secondary active transport has been suggested for the description of this situation. In an energy balance sheet as that presented by R.D. Keynes at this meeting, therefore, the energy for sugar transport would appear under the heading Na^+ transport. In how far the energy stores of Na^+ and K^+ gradients are sufficient to account for the "secondary active transport" alone, is at present discussed extensively in the case of amino acids, and a final answer appears not to have been reached yet. In the field of sugar transport less attention appears to have been devoted to this question. The problem, in how far sugar transport requires separate consideration in the discussion of energy costs, therefore, is largely open at present.

REFERENCES

1. Wilbrandt, W. (1969) Specific transport mechanisms in the erythrocyte membrane. Experientia 25, 673—677.
2. Wilbrandt, W. and Rosenberg, T. (1961) The concept of carrier transport and its corollaries in pharmacology. Pharmacol. Rev. 13, 109—183.
3. Wilbrandt, W. (1969) Carrier Systems in biological transport. Proc. 4th Int. Congr. Pharmacol. Basel. Vol. IV, pp. 391—401. Karger, Basel.
4. Rosenberg, T. and Wilbrandt, W. (1963) Carrier transport uphill. I. General. J. Theor. Biol. 5, 288—305.
5. Cohen, G.N. and Monod, J. (1957) Bacterial permeases. Bacteriol. Rev. 21, 169—194.
6. Kepes, A. (1970) Galactoside permease of Escherichia coli. In current topics in membranes and transport (Bronner, F. and Kleinzeller, A. eds), pp. 102—134, Academic Press, New York and London.
7. Winkler, H. and Wilson, T.H. (1966) The role of energy coupling in the transport of β-galactosides by Escherichia coli. J. Biol. Chem. 241, 2200—2211.
8. Kaback, H.R. (1972) Transport mechanisms in isolated bacterial cytoplasmic membrane vesicles. In the molecular basis of biological transport. (Woessner, Jr, J.F. and Huijing, F. eds), pp. 291—328, Academic Press, London and New York.
9. Riklis, E. and Quastel, J.H. (1958) Effects of metabolic inhibitors on potassium-stimulated glucose absorption by isolated surviving guinea pig intestine. Can. J. Biochem. Physiol. 36, 363—371.
10. Crane, R.K. (1962) Hypothesis for mechanism of intestinal active transport of sugars. Fed. Proc. 21, 891—895.
11. Schultz, S.G. and Curran, P.F. (1970) Coupled transport of sodium and organic solutes. Physiol. Rev. 50, 637—718.
12. Roseman, S. (1972) A Bacterial Phosphotransferase System and its Role in Sugar Transport. In the Molecular Basis of Biological Transport. (Woessner, Jr. J.F. and Huijing, F. eds), pp. 181—218, Academic Press, London and New York.
13. Mitchell, P. and Moyle, J. (1958) Group-translocation: a Consequence of enzyme-catalyzed group-transfer. Nature 182, 372—373.

Subject index

N-acetyldopamine, 44, 47
Actin, 131, 136, 142
Actinomycin, 121
Actomyosin, 150
Adrenaline, 223
Adrenergic control of blood flow, 223
Aequorea, 7
Alkylbenzenesulphonate, 223
Amino acids uptake, 245
Annelids, 143
Anthopleura, 7
Argyopelecus, 213
Asterias, 7
ATPase, 150, 174
ATPase activity, 64
ATPase inhibitors, 84
ATP utilisation, 244

Bacillus, 61
Bacteriophage, 67
Bioelectric control of ciliary activity, 75
Biomaterials formation of, 3
Biomaterials synthesis of, 4
Blood flow, adrenergic control of, 223
Byssus, 133

Ca^{2+}, 84, 91, 148
Ca^{2+} binding, 91, 103, 131
Ca^{2+} conductance, 75
Ca^{2+} regulation by, 143
Calliactis, 37
Carassius, 190
Carchesium, 83, 99
Ceratoscopelus, 213
Cichlasoma, 202
Ciliary activity bioelectric control of, 75
Clostridium, 66
Contractile system, 83
Crustacea, 143
Ctenopharyngodon, 190
Cuticle, 43
Cyprinodon, 201
Cyprinus, 190

Echinodermata, 7, 143
Ectreposebastes, 212

Elasticity, 26, 50, 52
Energy, 155, 161, 173, 187, 235, 243, 255, 259
Epipelagic fishes, 214
Epistylis, 83
Erythrocyte membranes, 21
Escherichia coli, 262

Flagella, 83
Flagella
   bacterial: structure and function, 61
Flagella
   motility of, 67
Flagella
   ultrastructure, 63
Food intake, 202
Food utilization, 188

Gadus, 228
Galactoside permease, 262
Glycolysis, 173

Holacanthus, 190
Hyperelastic materials, 9
Hyperelastic materials
   thermodynamics of, 10
Hyperelastic surface, 14

Ictalurus, 190
Insect cuticle, 43
Insect fibrillar muscle, 139
Ion pumping, 235
Ionic regulation, 161
Isoprenaline, 227

Kuhlia, 191

D-Lactate dehydrogenase, 262
Lebistes, 202
Lethocerus, 139
Lipid bilayer, 9
Locust, 44

Meganyctiphanes, 198
Membranes, 9, 75, 249
Membranes
   fluidity of, 16

Membranes
    mechanical instabilities in, 49
Membrane composite structures, 16
Mesopelagic fishes, 212, 214
Meromyosin, 105
Metridium, 7, 37
Micropterus, 188
Monolayers, 9
Motility, 67, 83
Muscles, 93, 105, 134, 211
Muscle contraction, 127, 139, 173
Myctophidae, 212, 213
Myctophum, 213
Myoblast fusion, 121
Myofibrils, 84, 141
Myogenesis, 121
Myosin, 105, 121, 136, 142, 150, 174
Mytilus, 133

Nerve membranes, 249
Nitella, 33
Noradrenaline, 223
Notropis, 188

Oncorhynchus, 189
Opalina, 78
Oryzias, 190
Oxidative phosphorylation, 187
Oxygen consumption, 192, 211

Pachynoda, 47
Paramecium, 75
Phosphotransferase, 263
Plasma membranes, 9
Protein, 105
Protein composition of the spasmoneme,
    99
Proteins
    structural, in relation to mechanical
    function, 43
Proteus, 65
Pseudomonas, 66

Rhodospirillum, 64
RNA synthesis, 121

Salmo, 190
Salmonella, 62
Salvelinus, 198
Sclerotization, 44
Scyliorhinus, 228
Serranus, 162
Solid state models, 26
Spasmoneme of vorticellids, 83, 99
Spirillum, 61, 68
Spirostomum, 84
Squid giant axon, 243
Stentor, 84
Sternoptychidae, 212, 213
Stress concentrations, 55
Structural systems, 3
Stylonychia, 76
Sugar transport, 259
Surface constriction, 15
Surface pressure, 15
Surface rigidity, 15
Surface tension, 10

Trachelocerca, 83
Trachurus, 212
Transport, 155, 163, 243, 249
Treponema, 61
Triphoturus, 213
Tropomyosin, 131
Teleost fishes
    bioenergetics of, 187
Tilapia, 193
Time-dependence, 6, 25

Viscoelastic models, 27
Viscosity, 26
Vorticella
    spasmoneme of, 83

Zoothamnium, 83, 99